T0224972

SpringerBriefs in Computer Science

SpringerBriefs present concise summaries of cutting-edge research and practical applications across a wide spectrum of fields. Featuring compact volumes of 50 to 125 pages, the series covers a range of content from professional to academic.

Typical topics might include:

- A timely report of state-of-the art analytical techniques
- A bridge between new research results, as published in journal articles, and a contextual literature review
- A snapshot of a hot or emerging topic
- An in-depth case study or clinical example
- A presentation of core concepts that students must understand in order to make independent contributions

Briefs allow authors to present their ideas and readers to absorb them with minimal time investment. Briefs will be published as part of Springer's eBook collection, with millions of users worldwide. In addition, Briefs will be available for individual print and electronic purchase. Briefs are characterized by fast, global electronic dissemination, standard publishing contracts, easy-to-use manuscript preparation and formatting guidelines, and expedited production schedules. We aim for publication 8–12 weeks after acceptance. Both solicited and unsolicited manuscripts are considered for publication in this series.

**Indexing: This series is indexed in Scopus, Ei-Compendex, and zbMATH **

Showmik Bhowmik

Document Layout Analysis

 Springer

Showmik Bhowmik
Department of Computer Science
and Engineering
Ghani Khan Choudhury Institute of
Engineering and Technology
Malda, West Bengal, India

ISSN 2191-5768   ISSN 2191-5776   (electronic)
SpringerBriefs in Computer Science
ISBN 978-981-99-4276-3   ISBN 978-981-99-4277-0   (eBook)
https://doi.org/10.1007/978-981-99-4277-0

This Springer imprint is published by the registered company Springer Nature Singapore Pte Ltd.
The registered company address is: 152 Beach Road, #21-01/04 Gateway East, Singapore 189721,
Singapore

Preface

Documents often possess information regarding the social, economic, educational, and cultural status of a particular place or person for a specific time period. Historical documents are the witnesses of the developments that have been made in various sectors of a country since the early days. Therefore, different archives are established around the world to preserve these documents in their physical form. However as these were created long years ago, there is a risk to preserve them in their physical form. These documents need to be digitized. Considering this fact, a widespread initiative of converting paper-based documents into electronic documents has been taken. There are lots of advantages of electronic documents over paper documents, such as compact and lossless storage, easy maintenance, efficient retrieval, and fast transmission. However, mere electronic conversion of these documents would not be of pronounced help for properly preserving the documents as well as automatic information retrieval, unless we provide a system for efficiently analysing the layout of these documents.

Documents have an explicit structure; it can be segregated into a hierarchy of physical modules, such as pages, columns, paragraphs, text lines, words, tables, and figures or a hierarchy of logical modules, such as titles, authors, affiliations, abstracts, and sections or both. This structural information would be very beneficial and convenient in indexing and retrieving the information contained in the documents.

The objective of document layout analysis is to detect these physical modules present in an input document image to facilitate effective indexing and information retrieval.

Malda, West Bengal, India   Showmik Bhowmik

Contents

Chapter 1
Introduction

Abstract Documents often carry crucial information, covering almost every aspect of human society. Even in this era of cyber-physical systems, many of us prefer to read paper documents. These facts raise the concern of careful storage and management of these documents. However, preserving these documents in their physical form has many risks and also limit their access. To address these problems, it is required to convert these paper-based documents into their electronic form. However, mere electronic conversion would not be of pronounced help for properly preserving the documents as well as automatic information retrieval. Therefore, a system for efficiently analyzing the layout of these documents becomes a pressing need. In this chapter, a quick introduction to Document layout analysis and its constituent stages are presented. This chapter also discusses various challenges associated with the task of layout analysis.

Keywords Document image processing · Document layout analysis · Binarization · Manhattan layout · Non-Manhattan layout · Overlapping layout · DIBCO · Document understanding

From ancient times documents have been used as an important medium of storing and conveying information. Even in this era of cyber-physical systems, many of us prefer to read paper documents. This common preference of human society causes a steady growth in the production of such documents from ancient times. Besides, every document generated for certain purposes irrespective of time always possesses important information. For example, historical documents often carry important historic information regarding a place, person, and event for a certain period. Whereas contemporary documents like books, magazines, reports, and other types cover the current economic, social and cultural status. Therefore, these documents should be carefully stored and managed. However, stockpiling of such ever-increasing paper documents is quite unwieldy. There is always a risk to preserve these in their physical form. These may get lost due to aging, casual handling, natural calamity, and many other reasons. Additionally, such arrangements cause limited access to these documents. For example, despite the importance of historical documents, for a long period, these documents are only found in their physical form in different libraries around the world.

To address these problems, it is required to convert these paper-based documents into their electronic form as there are lots of advantages of electronic documents over paper documents, such as compact and lossless storage, easy maintenance, easy access, efficient retrieval, and fast transmission, etc. Therefore, a widespread initiative has been taken to digitize these documents to store and make these available in their digital form. For example, such an initiative was taken by British Library in 2016 to digitize rare early Indian printed books, published between 1713 and 1947 [1]. British Library has digitized more than 3600 books so far and made these open through the project named *Two Centuries of Indian Print* [2]. Besides that, many research groups have also indulged themselves in digitizing contemporary documents to facilitate their research [3]. However, only digitizing paper documents by generating scanned images will not serve the purpose. The information present in these documents should be electronically available so that based on the need of the hour, this information must be used to facilitate many data mining applications. However, the conversion of such a voluminous document in its electronic form is not an easy task. Any attempt at manual conversion would be unrealistic due to their extensive volume. Thus, an efficient Document Image Processing (DIP) system which can automatically convert such paper documents into their electronic form has become a necessity.

1.1   Document Image Processing System

DIP transforms a scanned document image to an appropriate symbolic form [4]. A DIP consists of many sub-processes from different sub-disciplines which include, image processing, pattern recognition, natural language processing, artificial intelligence, and database systems. The output of a DIP system can vary based on its purpose. For example, the output can be a structural descriptor from which the entire document can be reconstructed, can be a semantic descriptor for performing task like sorting, filing, etc., or an editable file. Among these possible outcomes, output suitable for editing and reconstruction are popular.

A typical DIP system consists of following sub-processes:

- **Scanned image generation**: Scanned document images are the input to a typical DIP system. Generally, in this step document images are scanned using document scanners commonly in 300–400 dots per pixel (DPI). Two types of scanners are mostly used for the said purpose; Charged Coupled Device (CCD) scanners and Contact Imaging Sensor (CIS) scanners. CCD scanners use real lenses just like the modern cameras and capture high resolution images whereas, CIS scanners do not offer facility to generate high quality images. However, CIS scanners are less expensive. This subprocess of generating scanned images is often regarded as image acquisition step.

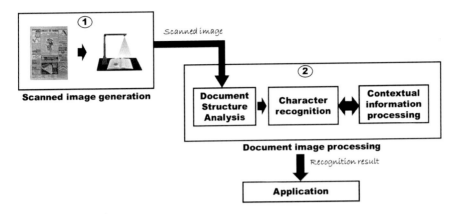

Fig. 1.1 Illustration of a typical DIP system

- **Document structure analysis**: Any document image consists of several regions, which can be of various types like paragraphs, images, tables, etc. Paragraphs contain lines, lines are made of words, words possess characters. To understand the structure of a document it is necessary to decompose it into its constituent elements and identify their spatial relationship.
- **Recognition**: Recognition of characters is necessary to produce required outputs. To do that often contextual information is used. The reason for that in documents like unconstrained handwritten or historical printed, recognition of character in isolation is very difficult.

Diagram of a typical DIP system is given in Fig. 1.1.

1.2 Structure of a Document

A document image is a collection of several physical entities that include text blocks/paragraphs, text lines, words, tables, charts, figures along background. Kise in his survey [5], presents a diagram indicating the hierarchical structure of a document image where a document is regarded as a collection of pages and pages contain several blocks like text, figure, table, etc. Text blocks contain text lines, lines contain characters, and so on. This structural information would be very beneficial and convenient in indexing and retrieving the information contained in the document, thereby signifying the need for document structure analysis.

The spatial arrangement of different constituent regions and their relationship of a document can be considered as the layout or structure of this document. Haralick in his work [6] has regarded that documents have both *geometric* and *logical struc-tures*. As per the author, these structures can be defined as follows:

- **Geometric structure** *of a document image* $\varnothing = (R, S)$, *where* R *is the set of regions and* S *is a labeled special relation on* R. *A region* $R = (T, \theta)$, *here* T *signifies the geometric type of the region (i.e., circle, ellipse, square, rectangle or polygon) and* θ *is the set of corresponding parameter values. The labeled spatial relation* $S \subseteq R \times R \times L$ *contains triplets* (R, R', l) *which signifies the region* R *stands in the relation* l *with the region* R'. l *can be atomic, inside of, mutually exclusive of, overlapping with, or any other complex relation.*
- **The logical structure** *of a document includes the functional labels of different constituent regions and the reading order of the text regions.*

1.3   Categories of Document Layout

In the early days of document image processing, DLA was not considered as a complete and complex research problem, rather just a pre-processing step having some minor challenges [7]. The main reason for that is the type of layout being considered for processing was simple. Researchers are paying attention to this complex problem as they come across a large variety of documents. Based on the complexity of the layouts, it can be divided into different categories [5]. The geometrical structure or layout of a document can vary widely. Kise in his work [5] has classified the document layouts *as Rectangular, Manhattan, Non-Manhattan, and Overlapping* considering the text block-level. *The rectangular layout* is regarded as the most fundamental layout. Here all the regions' borders can be represented by non-overlapping rectangles whose boundaries are either parallel or perpendicular to the page boundaries (see Fig. 1.2a). Generally, the pages of business documents and books follow this type of layout. However, in some cases, the region boundaries possess a concave nature. Therefore, boundary representation with a non-overlapping rectangle is not possible (see Fig. 1.2b). According to the author, if the region boundaries are sided parallel or perpendicular to each other as shown in Fig. 1.2b then it is called *Manhattan layout*. This type of layout is often

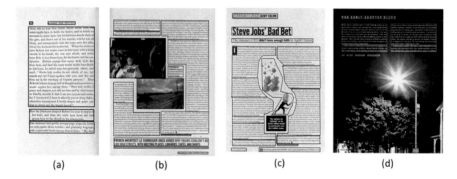

(a)	(b)	(c)	(d)

Fig. 1.2 Different layouts (**a**) rectangular, (**b**) Manhattan, (**c**) non-Manhattan, and (**d**) overlapping

found in newspapers, magazines, etc. *Manhattan layout* includes *Rectangular layout* as its sub-class. The layout in which the region boundaries are non-overlapping but do not hold parallel or perpendicular sides is known as *the Non-Manhattan layout* (see Fig. 1.2c). Generally, pages of magazines with large figures possess this type of layout. Although most of the printed documents follow any of these said layouts, there are certain documents like book covers, magazine covers, graphical magazines, etc. where text can appear on a complex and colorful background or image (see Fig. 1.2d) or two text lines intersect each other or get intersected by an image do not come under the purview of the said non-overlapping layouts. These types of layouts are called *Overlapping layouts*.

Besides the said classification, Kise [5] has also classified the document layouts from the text line perspective i.e., *Horizontal* and *vertical text lines, Parallel text lines,* and *Arbitrary text lines.* In the first category, text lines appear parallel to either the horizontal or vertical side of the page. This is the most common and fundamental layout. On the other hand, *Parallel text lines* are only parallel to each other, but not horizontally or vertically aligned. This type of layout is relatively rare and mostly found in advertisements. *Arbitrary text lines* are very uncommon in printed documents. However, it can be found in manuscripts. In this type of layout, text lines can follow any arbitrary shape.

It is to be noted that all these layouts are discussed for non-skewed images. This classification of layouts is mostly used to categorize and analyze the methods available in the literature for DLA. Many researchers have used the concept of layout to classify documents in different complexity classes. For example, Okun et al. in [8] classify documents in five classes based on their layout which are *Structured articles, Unstructured layout, Semi-structured layout, maps and engineering drawing,* and *non-traditional documents.* In the first category, the documents having regions normally physically separated from each other are kept. However, the authors have also mentioned that the gaps between these regions can be very narrow. The articles published in journals, newspapers, and newsletters are considered in this category. In the second category, authors have placed advertisements, cover, or title pages of CDs, books, and journals. The layout of these documents is regarded as unstructured as these do not follow any general rule and depend upon the designer's goal. The third category i.e., semi-structured layout possesses few structural elements. However, there is no specific location defined in the document pages for these elements. Envelopes, postcards, business cards, bank checks, forms, and table-like documents are placed in this category. In the fourth category, maps and engineering drawings are considered. Unlike the other documents, the components of these documents appear in a scattered way. These documents contain sparse text among the graphical components with very large dimensions. Besides that, maps often possess many color levels indicating certain vital information. In the last category, web documents, and video frames are kept. These documents are mostly in color and low resolutions. In these documents, images and text can appear arbitrarily.

Similar to this classification, recently Bhowmik et al. in [9], have initially classified the document images in four broad categories: *Online document images,*

Scene images and Video frames, Web images, and *General offline document images,* and then further classified the *General offline document images* into four complexity classes. Although these classifications are done from the aspect of text non-text separation but these classifications are indirectly influenced by the structural complexity.

1.4 Document Layout Analysis (DLA)

Tang et al., in [10], have described a basic model of document processing, where the process of extracting the geometrical structure of a document image is regarded as *Document analysis* or *layout analysis* and mapping the geometric structure to logical structure is stated as *Document understanding* (see Fig. 1.3). However, the authors have also mentioned that it is often impossible to draw a clear boundary between these. In [5], Doermann and Tombre have described document layout analysis (DLA) as the process of segmenting an input document image into various homogeneous components like, text blocks, figures, tables, etc., and classifying them accordingly. Similarly, Binmakhashen et al., in [11], have described DLA as the process of detecting and annotating the geometric structure of an input document image.

Figure 1.4 shows sample input and expected output of a DLA system, where in the output image text regions are represented using blue outlines which include paragraphs and headlines. The regions with pink outlines are separators. Regions (rectangular) with dark green outlines are images and the ones with light green outlines are graphics.

A complete DLA system consists of various elementary phases. Due to the variations in document layout types, the phases of a DLA system may vary largely. However, by observing the literature, two most common phases can be found which

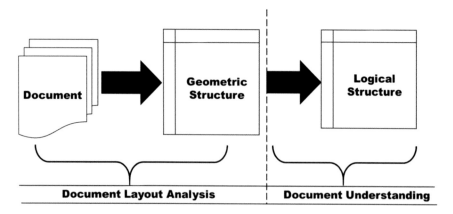

Fig. 1.3 Illustration of the document processing model

(a) Input document image [10]

(b) Expected output [13]

Fig. 1.4 Illustration of (**a**) input and (**b**) the corresponding output of a typical DLA system for printed documents

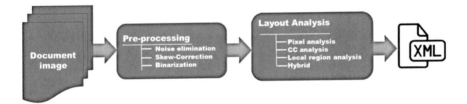

Fig. 1.5 Framework of a typical DLA system

are *Pre-processing* and *Layout Analysis*. Some methods also perform certain *post-processing* [9]. A general framework for a DLA system is given in Fig. 1.5.

- **Preprocessing**: The *Layout Analysis* phase of any DLA method often makes certain assumptions about the input images like input images will be noise-free, binarized, with no skew or all. The objective of the pre-processing step is to transform the input image as per the requirement of the layout analysis phase. More detailed discussion on pre-processing is given in Chap. 2.
- **Layout analysis**: Analysis of layout involves identifying the boundaries and types of the constituent regions of an input document image. The processes of identifying the region boundaries is called *document region segmentation* and

classifying the identified regions as per their type is called *document region classification*. In Chap. 3, the region segmentation process and in Chap. 4, the region classification process are discussed in detail along with the state-of-the-art methods.

- **Final representation**: Finally, both the extracted geometrical and logical structures are to be stored for reconstruction. For that generally the hierarchical tree data structure is used. One popular way of storing such tree structure is XML files.

1.5   Why DLA Is Still an Open Area of Research

Before discussing the research gaps, it is required to understand the challenges in DLA and how far these have been addressed. As discussed earlier, a DLA system includes identifying the exact cut points to generate the segmented regions. It is easier to develop such a method for a specific type of layout. However, it is very difficult to design a generalized method that can do the same for a wide range of layouts. This implies, when the layout is strictly constrained, fixed cut points will serve the purpose. Whereas, when the layout is completely unconstrained, it is necessary to understand the content. That defines the range of challenges in developing a DLA system to be useful for a practical scenario.

Another aspect of this is using a very complex method for a relatively restricted layout would be unnecessary and may cause loss of efficiency and accuracy. Whereas applying a less generalized method to a layout beyond its capability may cause a disaster. Therefore, it poses another challenge that is to know the class of layout before processing.

Research on DLA has been started long back, and several methods have been introduced by researchers for that purpose. However, the methods reported earlier are capable of handling the layouts up to Manhattan quite efficiently [5]. But their applicability gets limited for the layouts beyond Manhattan. In recent days, the continuous advancement of multimedia technology causes the generation of more complex and decorative paper documents. As a result, documents are becoming more expressive to the readers and able to attract their attention as well [9]. This in turn makes the layouts more and more free-form or less constrained. Therefore, it is required to design methods to deal with these documents having complex layouts along with the simple ones. Considering this fact, the ICDAR community is arranging different competitions since 2001 [12–14] on DLA to address the related challenges in the contemporary documents. From careful observation, it can be noticed that the number of participating methods has increased in the recent competitions compared to the earlier versions (see Fig. 1.6). That indicates the problem of layout analysis is still an open area of research.

In contrast to the contemporary documents, the layouts of the historical printed documents are relatively constrained. However, the historical documents suffer from many quality level degradations. Binmakhashen et al. in [11] have mentioned two

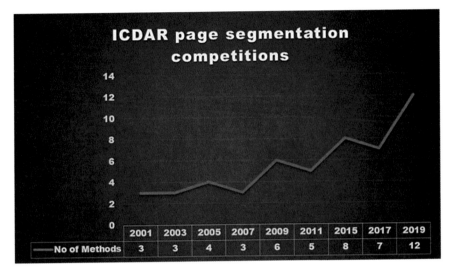

Fig. 1.6 Illustrates the number of participating methods in different page segmentation competitions organized by ICDAR

types of degradations that can affect the performance of a DLA method i.e., native, and auxiliary. The first category of degradations appears due to aging, casual handling, the ink used, paper quality, writing style, etc. Although for printed documents the last cause is not applicable, the font style and font size also vary for this type of documents. See-through, ink-seepage, text fading, etc. come under the purview of this category. The degradations from the second category appear due to glitches in the scanning device, alignment during scanning, lighting condition during acquisition. The issues that come under this category are skew, blurring, dark spots, etc. Therefore, pre-processing plays a vital role while dealing with historical documents. From the literature survey, it can be observed that a large portion of methods that are made available for DLA consider binarization in their pre-processing stage. However, most of these methods have less emphasized on their pre-processing step. Many of these methods have also assumed non-skewed input images [11]. Therefore, an efficient pre-processing method is required to aid a DLA method for proficiently segmenting a historical document image. Although some deep learning-based methods are introduced recently which do not need a binarized image as input, these methods require huge amounts of training data. Therefore, in absence of sufficient labeled historical data, these methods fail to perform at per [15].

Research on binarization was started long back. However, an efficient method that can handle most of the native degradations is yet to be built. For example, in a recent binarization competition DIBCO 2019 [16], 24 methods from different research groups were submitted. Out of 24 methods, only 3 methods have obtained more than 70% F-measure. This indicates that in addition to the development of an efficient DLA method, the design of an efficient pre-processing method to suppress the degradations at the initial level is a pressing need.

References

1. Clausner, C., Antonacopoulos, A., Derrick, T., Pletschacher, S.: ICDAR2017 Competition on Recognition of Early Indian Printed Documents—REID2017. http://usir.salford.ac.uk/id/eprint/44370/1/PID4978585.pdf (2017)
2. The British Library. Two Centuries of Indian Print. https://www.bl.uk/projects/two-centuries-of-indian-print
3. Antonacopoulos, A., Bridson, D., Papadopoulos, C., Pletschacher, S.: A realistic dataset for performance evaluation of document layout analysis. In: 10th International Conference on Document Analysis and Recognition, 2009 (ICDAR'09), pp. 296–300 (2009)
4. Srihari, S.N., Niyogi, D.: Document image processing. Wiley Encycl. Electr. Electron. Eng. (2001)
5. Kise, K.: Page segmentation techniques in document analysis. In: Handbook of Document Image Processing and Recognition, pp. 135–175. Springer (2014)
6. Haralick, R.M.: Document image understanding: geometric and logical layout. In: CVPR, vol. 94, pp. 385–390 (1994)
7. Wong, K.Y., Casey, R.G., Wahl, F.M.: Document analysis system. IBM J. Res. Dev. 26(6), 647–656 (1982)
8. Okun, O., Dœrmann, D., Pietikainen, M.: Page segmentation and zone classification: the state of the art. DTIC Document (1999)
9. Bhowmik, S., Sarkar, R., Nasipuri, M., Doermann, D.: Text and non-text separation in offline document images: a survey. Int. J. Doc. Anal. Recognit. 21(1–2), 1–20 (2018)
10. Tang, Y.Y., Cheriet, M., Liu, J., Said, J.N., Suen, C.Y.: Document analysis and recognition by computers. In: Handbook of Pattern Recognition and Computer Vision, pp. 579–612. World Scientific (1999)
11. Binmakhashen, G.M., Mahmoud, S.A.: Document layout analysis: a comprehensive survey. ACM Comput. Surv. 52(6), 1–36 (2019)
12. Gatos, B., Mantzaris, S.L., Antonacopoulos, A.: First international newspaper segmentation contest. In: Proceedings of Sixth International Conference on Document Analysis and Recognition, pp. 1190–1194 (2001)
13. Clausner, C., Antonacopoulos, A., Pletschacher, S.: ICDAR2017 competition on recognition of documents with complex layouts—RDCL2017. In: 2017 14th IAPR International Conference on Document Analysis and Recognition (ICDAR), vol. 1, pp. 1404–1410 (2017)
14. Clausner, C., Antonacopoulos, A., Pletschacher, S.: ICDAR2019 competition on recognition of documents with complex layouts—RDCL2019. In: 2019 International Conference on Document Analysis and Recognition (ICDAR), pp. 1521–1526 (2019)
15. Clausner, C., Antonacopoulos, A., Derrick, T., Pletschacher, S.: ICDAR2019 competition on recognition of early Indian printed documents—REID2019. In: 2019 International Conference on Document Analysis and Recognition (ICDAR), pp. 1527–1532 (2019)
16. Pratikakis, I., Zagoris, K., Karagiannis, X., Tsochatzidis, L., Mondal, T., Marthot-Santaniello, I.: ICDAR 2019 competition on document image binarization (DIBCO 2019). In: International Conference on Document Analysis and Recognition (ICDAR), pp. 1547–1556 (2019)

Chapter 2
Document Image Binarization

Abstract Documents often get affected by noise due to various issues like casual handling occurred during storing, or image acquisition processes. These noises not only affect the quality of the documents but also degrade the visual appearance of the content. Therefore, these have high potential to downgrade the final outcome. To suppress the noise, occur in an input document image at the initial stage, it is essential to use an efficient pre-processing before further analysis. Pre-processing converts the input image to a specific form, suitable for further analysis without disturbing the knowledge content. Among all, binarization is very popular and commonly performed pre-processing before many document image processing tasks. Therefore, in the present chapter, a detailed discussion on document image binarization and the recent progresses in this domain are presented. Additionally, in this chapter, various degradations related to document image, and various noise models are also described to make the discussion complete.

Keywords Binarization · Noise · Bleed through · Fade ink · Pre-processing · Deep learning · Shallow learning · DIBCO

Documents often suffer from many types of degradations for various reasons. For example, due to aging, ink usage, or poor quality of paper, documents may suffer from issues like show-through, ink-bleeding, text fading, touching text, etc. In addition to these, document images may also get noise during image acquisition or scanning process. Presence of noise affect the quality of the final output negatively. Therefore, it becomes essential to suppress the noise present in the input images before further analysis. Despite its importance, most of the DLA methods available in the literature have not emphasized on their pre-processing stage. But to make the discussion complete, it is necessary to have a discussing on document image pre-processing.

© The Author(s), under exclusive license to Springer Nature Singapore Pte Ltd. 2023
S. Bhowmik, *Document Layout Analysis*, SpringerBriefs in Computer Science,
https://doi.org/10.1007/978-981-99-4277-0_2

2.1   Different Types of Degradations and Noise

One of the major objectives of pre-processing is to suppress the noise and deal with
the degradation at the early stage, so that further analysis can be effectively done.
The degradation refers to the loss of stored information in the image whereas noise
represent the unwanted information added from various sources to the image. Any
degraded image $D_{(x, y)}$ can be mathematically defined as,

$$D_{(x,y)} = I_{(x,y)} * d_{(x,y)} + \lambda_{(x,y)} \tag{2.1}$$

where, $I_{(x, y)}$ is the original image, $d_{(x, y)}$ is the degradation and $\lambda_{(x, y)}$ is the noise. In
this equation '*' refers to 2D convolution operation.

Noise in document images may appear due to some inherent issues or during
acquisition due to external conditions. Therefore, noise sources of a document image
can broadly be categorized as *Native* and *Auxiliary* [1]. Noise from native sources
appear due to aging, quality of paper and ink used. It causes issues like,

- **Ink-bleeding or bleed through** appears when both sides of the page are used to
  write and the ink seeps through one side and spreads over to the other side (see
  Fig. 2.1a, b).
- **Show through or smear** appears because of the ink impression of one side
  appears on other side. It often creates dark spots on the page (see Fig. 2.1c, d).

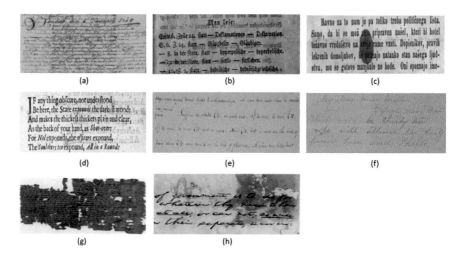

Fig. 2.1 Presents document images with noise from native sources, (**a, b**) sample document images
from Document image binarization competitions (DIBCO) dataset with bleed through/ink-bleed
degradation [13], (**c, d**) sample document images from DIBCO dataset with show through/smear
degradation [5], (**e, f**) sample document images with faint text degradation [5], and (**g, h**) sample
document images from DIBCO dataset with issues due to document deterioration [14]

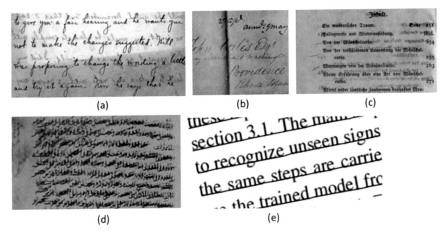

Fig. 2.2 Presents document images with noise from auxiliary sources (**a**) sample document image from DIBCO dataset with uneven illumination, (**b**) sample document image from DIBCO dataset with contrast variation issue (low contrast at middle and high contrast at both sides), (**c**) sample document image with out-of-focus blurring effect [5], (**d**) sample document image with motion blurring effect [5], and (**e**) sample image with skew

- **Faint text** appears in documents written or printed using low quality ink. This is because as the time grows the ink starts to shrink. The quality of paper used may also be the reason for that (see Fig. 2.1e, f).
- **Deterioration of document** takes place due to aging, poor storage, mishandling, natural calamity and other environmental condition. This type of noise causes presence of dark spot and other artefacts on the document images (see Fig. 2.1g, h).

In addition to the noise from native sources, document images also get noise due to many external factors (or from auxiliary sources) like malfunctioning of scanning devices, lighting condition during acquisition, misplacement of the document during scanning, etc. These include,

- **Uneven illumination** occurs in the light microscopy images as in optical imaging, the incident light decreases drastically along the path [2]. This is because of the scattering of particles in the media [3]. Additionally, due to the background object, superposition of fluorescence absorption and emission spectra often light gets disseminated which also results in uneven illumination [2] (see Fig. 2.2a).
- **Contrast variation** appears due to the environment under which the image acquisition process is carried out. For example, sunlight, illumination of light, obstruction often results in contrast variation in images [4] (see Fig. 2.2b).
- **Blurring effects** observed in document images are of two types; motion blurring and out-of-focus blurring. Motion blurring appears due to the relative motion of the camera and object or sudden rapid movement of the camera (see Fig. 2.2d). Out-of-focus blurring appears when some of the points are in focus while others are not during acquisition [5] (see Fig. 2.2c).

- **Skew** in document images appears either due to the misplacement of the document during scanning or due to the writing style. This can occur either at the page level or at the region level or both. Generally, for the printed documents, page level skew is found whereas in handwritten documents region level skew is observed mostly.

In addition to the said types of noise observed in document images, few other types of noise are discussed in this section for the sake of completeness. During transmission, acquisition often, the digital data get modified due to the addition of some unwanted information. In terms of signal processing, some unwanted signals may get added to the desired one which results in random intensity variation in the spatial domain. To measure the amount of noise got added and develop suitable restoration technique various noise models are developed. These include,

- **Gaussian noise** often known as electronic noise as it appears in the amplifier at the receiving end [6]. The reason for occurring Gaussian noise is thermal vibration of atoms. The probability density function for Gaussian noise can be expressed as,

$$p(i) = \sqrt{\frac{1}{2\Pi\sigma^2}} e^{-\frac{(i-\mu)^2}{2\sigma^2}} \qquad (2.2)$$

where, i is the intensity from the original image, μ represents mean, and σ is standard deviation.

- **White noise** normally modifies the image pixel values positively as the autocorrelation in white noise is zero [7]. Besides, noise power spectrum for white noise is constant. Noise power refers to the total amount of noise in a bandwidth measured at the output or input of a device in absence of signal [6].
- **Brownian noise** appears due to random movement of suspended particles in fluid which results in Brownian motion. It can also be generated from white noise [6].
- **Salt and pepper noise** also known as drop noise or impulse noise. In this noise, instead of entire image some pixel values get modified. That means even after corrupted by salt and pepper noise some pixels in effected image remain same as the original [6, 8]. The probability density function for Salt and pepper noise can be expressed as

$$p(i) = \begin{cases} p_a & if \ i = a \\ p_b & if \ i = b \\ 0 & otherwise \end{cases} \qquad (2.3)$$

where, i is the intensity from the original image, b is a brighter point if $b > a$, and b is darker if $b < a$.

- **Periodic noise** appears due to electronic interferences during image acquisition. This noise is sinusoidal in nature and spatially dependent [9].

- **Speckle noise** appears in the imaging systems like radar, laser acoustics, etc. This is a multiplicative noise and appears in images as Gaussian noise. The probability density function of Speckle noise follows gamma distribution [6].
- **Poisson noise** appears in images generated using x-ray, gamma ray imaging systems. This noise is also known as quantum noise or shot noise. The probability density function of this noise follows Poisson distribution [6].
- **Poisson Gaussian noise** appears in the images generated from Magnetic Resonance Imaging (MRI) system [6].
- **Structured noise** is also known as rank noise. It appears due to the interference of electronic components. This noise can be periodic, aperiodic, stationary or non-stationary in nature [6].
- **Gamma noise** also commonly appears in laser based images. The probability density function for gamma noise follows gamma distribution [10].
- **Rayleigh noise** is mostly found in radar images.

A more detailed discussion on different noise models can be found in [11, 12].

2.2 Pre-processing

Pre-processing is a process of transforming the input image to a specific form, suitable for further analysis. This step does not aim to modify the knowledge content of the input image but facilitates to extract. Some common and popular pre-processing operations before document image analysis are conversion of color image to grayscale, noise removal, binarization, skew correction and many more. However, the pre-processing may vary according to the targeted analysis of the input image. For example, layout analysis phase of DLA methods often make certain assumptions about the input images like it will be noise-free and binarized, with no skew or all.

Binarization is the processes of separating foreground pixels from the background pixels, so that every pixel will contain either 0 (foreground pixel) or 1 (background pixel) as intensity value (see Fig. 2.3). Whereas presence of skew refers to the

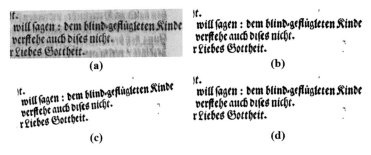

Fig. 2.3 Presents (**a**) original image, (**b**) corresponding binarized image, (**c**) image with skew and (**d**) corresponding skew corrected image

degradation where the content of a document image got rotated at a slight angle. Skew in a document image appears at the page-level as well as the region-level. In printed documents, mostly, the page-level skew is found (see Fig. 2.3). The presence of skew in an input document image can affect the final segmentation result. Therefore, it is essential to keep the input document image in $0°$ skew format before extracting the region information. Among these, binarization is very popular and commonly performed before many document image processing tasks. Therefore, in the present chapter, a detailed discussion on document image binarization and the recent progresses in this domain are presented.

2.3 Document Image Binarization

Binarization converts a grayscale image into a binary image where each pixel contains either 0 or 1 as the value. It can formally be defined as the function B_{in} such that

$$B_{in} : I_g \rightarrow I_b \qquad (2.4)$$

where $I_g = \{i \mid i \in \mathbb{Z} \text{ and } 0 \leq i \leq 255\}$, and $I_b = \{i \mid i \in \{0, 1\}\}$.

Document image binarization (DIB) is comparably the most popular and important step in the document layout analysis pipeline. A large portion of the layout analysis methods available in the literature considers binarization as the pre-processing step [15]. Especially, the methods that are developed to deal with historical documents are relatively more dependent on the binarization step. Due to its wide range of applications, DIB is one of the most popular research topics in the domain of image processing [14, 16, 17].

However, the binarization of degraded document images is a very challenging task. As discussed in Sect. 2.1, degraded document images suffer from various degradations and noise which pose the difficulties in the way of binarization. For example, the presence of stains, artifacts, creases in the degraded document images due to deterioration minimize the contrast between the foreground and background pixels inside the effected region. The sharp change of contrast around the boundary of the effected region also hard to identify separately from the normal foreground background boundary. Bleed through is most complex type of degradation found in document images from DIB perspective due to its text like shape. The complex background due to uneven illumination, contrast variation, document texture create difficulty for the extraction of statistical features as statistics get changes due to the intensity variation of regions. Presence of faint text makes the binarization challenging due to low contrast. The show through or smear issue creates problem in localizing the foreground region. Because of the verity of degradations, it is hard to design a common DIB method which will be equally effective for every image [18].

2.4 Different Binarization Methods

In this section, different methods developed for document image binarization are discussed. The DIB methods introduced in the literature can broadly be categorized into three major groups: *Threshold-based methods, optimization-based methods,* and *classification-based methods.*

2.4.1 Threshold-Based Methods

These are the classic solutions to the binarization problem and these were the first to appear in the literature. These methods generally compute a threshold T_h to classify the pixels in the input gray scale image I_g as foreground or background to generate the binarized image I_b as,

$$I_b(x, y) = \begin{cases} 0 & if\ I_g(x, y) < T_h \\ 1 & otherwise \end{cases} \tag{2.5}$$

where, $I_g(x, y)$, and $I_b(x, y)$ represent the pixel value at the (x, y) coordinate. Depending on the approach of calculating T_h, we can further divide these methods as (1) *global methods* [19, 20] and (2) *local methods* [21, 22].

Global methods compute a single threshold for the entire image, typically based on global statistics. Otsu [19] introduces the global threshold based methods which becomes very popular. This method considers the histogram h of the input image and a temporary threshold t_h such that $o \leq t_h \leq 256$. Every value of t_h divides the histogram h in two clusters. The final threshold T_h is set the value of t_h which minimize Eq. (2.6) and maximize Eq. (2.7).

$$T_h = \operatorname*{argmin}_{t_h} \left\{ W_{c1}(t_h)\sigma_{c1}^2(t_h) + W_{c2}(t_h)\sigma_{c2}^2(t_h) \right\} \tag{2.6}$$

$$T_h = \operatorname*{argmax}_{t_h} \left\{ W_{c1}(t_h)W_{c2}(t_h)(\mu_{c1}(t_h) - \mu_{c2}(t_h))^2 \right\} \tag{2.7}$$

where, $\quad W_{c1}(t_h) = \sum_{i=0}^{t_h} h(i), \quad W_{c2}(t_h) = \sum_{i=t_h}^{255} h(i), \quad \mu_{c1}(t_h) = 1/W_{c1} \sum_{i=0}^{t_h} h(i),$

$\mu_{c2}(t_h) = 1/W_{c2} \sum_{i=t_h}^{255} h(i), \quad \sigma_{c1}^2(t_h) = 1/W_{c1} \sum_{i=0}^{t_h} h(i)(i - \mu_{c1}(t_h))^2, \quad$ and $\quad \sigma_{c2}^2(t_h) =$

$1/W_{c2} \sum_{i=t_h}^{255} h(i)(i - \mu_{c2}(t_h))^2.$

Otsu method is very effective when the image histogram has two clear peaks. Besides, it has no parameter to be tuned.

Unlike global methods local methods estimate the local statistics in a close neighborhood of pixels to determine the threshold for that particular local area. *Niblack* [21] proposes a simple adaptive local method of binarization. In this method, a threshold is computed for each pixel (x, y) by estimating the local mean μ_L and variance σ_L^2 within the close neighborhood of the pixel under consideration which can be expressed as follows

$$T_h(x, y) = \mu_L(x, y) + k\sigma_L^2(x, y) \tag{2.8}$$

Here, k is the parameter to be tuned. The recommended vale of k is -0.2. *Sauvola and Pietikäinen* [23] introduce a variant of the Niblack's method as,

$$T_h(x, y) = \mu_L(x, y)\left[1 + k\left(\frac{\sigma_L^2(x, y)}{R} - 1\right)\right] \tag{2.9}$$

Here, the recommended value of k is 0.5. R is a constant set to the maximum possible standard deviation. An extension of this method is presented in [24] where the local statistics are normalized using the global one as

$$T_h(x, y) = \mu_L(x, y) - k\left(1 - \frac{\sigma_L^2(x, y)}{S_G}\right)(\mu_L(x, y) - M_G) \tag{2.10}$$

where, $S_G = \max\limits_{(x, y)}\left\{\sigma_L^2(x, y)\right\}$, and $M_G = \min\limits_{(x, y)}\left\{\mu_L(x, y)\right\}$. This method works effectively in limited contrast and limited intensity set. In general, global methods are relatively fast and found useful when a stable intensity level difference is present between foreground and background pixels but they are less effective for documents with uneven illumination, patches, and other degradations. In such situations, local methods are found effective compare to the global methods [25]. For example, in [26], local methods are shown to work well compared to global methods in slowly changing backgrounds. Similarly, in [27], the authors consider 40 binarization methods, including both global and local methods, to evaluate their performance for noisy document images. They also reach to the same conclusion. However, a major issue with these local methods is that they require a set of parameters to be tuned, and the performance of these methods is often sensitive to these parameter values [28]. However, for documents with complex backgrounds, a set of generalized parameter values that can perform equally well for different kinds of images is difficult to find. Considering this, in some recent literature [29, 30], researchers have added denoising or contrast-enhancement step before applying a local or adaptive technique. This additional pre-processing step helps in handling the uncertainty present in the adaptive thresholding methods with a generalized set of parameter values for extremely noisy document images.

2.4.2   *Optimization Based Methods*

These methods have arrived later compares to the first category. Most of these binarization methods use Conditional random field (CRF) [31, 32] which considers the spatial dependencies of pixels for binarization. To do that every CRF method follows a cost or energy function. The objective of this energy function is to indicate effectiveness of the binarization with respect to the input image. A typical cost or energy function used in these methods can be expressed as

$$E_{CRF}\left(I_b, I_g\right) = \sum_{x=1}^{N} E_x\left(I_b^x, I_g\right) + \sum_{x=1}^{N}\sum_{y=1}^{x} E_{(x,y)}\left(I_b^x, I_b^y, I_g\right) \tag{2.11}$$

Here, I_g represents input image, I_b binarized image, E_x and $E_{(x,y)}$ are energy functions defined over single and pair of pixels respectively. As said $E_{CRF}(I_g, I_b)$ indicates the goodness of the results which needs to be minimized for effective binarization as

$$\bar{I}_B = arg \min_{I_b} \left\{E_{CRF}\left(I_b, I_g\right)\right\} \tag{2.12}$$

The exact solution to Eq. (2.12) can only be reach for binarization if $E_{(x,y)}$ satisfies $\forall_{(x,y)}\{E_{(x,y)}(0,0,I_g) + E_{(x,y)}(1,1,I_g) \leq E_{(x,y)}(0,1,I_g) + E_{(x,y)}(1,0,I_g)\}$. In that case, E_{CRF} can be reformed as a weighted graph with positive weights and the minimum cut of this graph will be $\bar{I}_B$ [33]. In, [31] authors use CRF model for DIB where E_x is set to the image Laplacian, and $E_{(x,y)}$ is 0 for uniform regions. For the estimation of $E_{(x,y)}$, authors perform edge detection which is later modified [34]. This method [31] became very popular and secures first position in DIBCO binarization competition 2012 [35], 2014 [36], 2016 [37], and 2018. In [38], a CRF model is used where $E_{(x,y)}$ is estimated from the distances between background, and foreground pixels, and the skeleton image of the initial binarization result. Authors in [39] initially classify pixels into background, foreground, and uncertain before applying CRF. Here the CRF model is used to classify the pixels which were labeled as uncertain initially.

In [40] the researchers segment the input image into text, near text, and background zones and then apply a graph-cut algorithm to generate the final binarization result. In [41], simulated annealing (SA) is applied to minimize the cost function. Most of these methods use stroke width information to binarize document images and achieve good results for printed documents. However, for handwritten documents, these methods may not be as accurate in the estimation of the stroke width. Despite of these, the method reported in [31] gain popularity by winning four constitutive DIBCO competitions. Optimization-based methods are relatively time-consuming compared to the earlier category of methods.

2.4.3 Classification-Based Methods

Methods from this category are comparably new in the domain of document image binarization. Here, first, the *classical machine learning* based binarization methods are presented then *deep learning* based methods are discussed.

Classical machine learning based binarization methods include both supervised and unsupervised learning based approach. One such method is proposed in [42], where the authors develop a neuro-fuzzy technique for binarization and gray value (or color) reduction of poorly illuminated documents. In [43], authors use Multi-layer perceptron (MLP) to binarize Brazilian checks with complex background. To classify each pixel, the pixel intensity and mean intensities from four different 3×3 windows around the pixel under consideration are estimated as feature. In [44], the global intensity mean and standard deviation are considered along with all pixel intensities within a 3×3 window to represent the center pixel at the feature space. Finally, an MLP is used to classify the pixels based on the extracted features. Due to the restricted context, these models fail to generalize their learning over a wide variety of documents. Authors in [45], initially, estimate the background of the input image. Then to classify any pixel, position a 9×9 window on it by keeping the pixel under consideration at the middle and take all 81 intensities as feature. Additionally, position a 3×3 window in the estimated background image in the corresponding location and extract all 9 intensities as feature. These 90 features are then used to train an MLP for the classification. In this work, authors also device a loss function which is a continuous version of the f-measure and used during the training of the MLP. It is observed that the generalization capability of the model is improved due to the improved training.

Some studies [26, 46] suggest that developing a common binarization technique for handling various types of noise would be difficult. Thus, the authors in [47, 48] combine several binarization techniques using a Kohonen self-organizing map (KSOM). The combined method outperforms other binarization methods in most cases. In [49], mixture of Gaussian distributions is used to cluster the pixels of a document image based on local information. In [50], authors follow quad-tree-based segregation of the normalized grayscale image and then cluster the pixels in each partition using K means. The clustering is done based on the locally extracted features. A similar approach is reported in [51] but using C-Means algorithm. However, the performance of the said clustering algorithms heavily depends on the initial cluster centers, and therefore random initialization may lead to a situation where the produced clustering result may vary over different runs, which is not desirable. To address this issue, in a recent work [52], researchers have designed an initialization method to select the initial cluster centers before using the K-means algorithm for binarization. Here, to extract the local features, authors have designed a two-player, non-zero-sum, non-cooperative game at the pixel level. The extracted features for each pixel are then inputted to the K-means algorithm to classify a pixel as foreground or background.

Deep learning based methods become the state-of-the-art in the domain of binarization. In DIBCO 2017 [13], and 2019 [14] a significant number of deep learning based methods were submitted and the methods secure the top 6 positions in DIBCO 2017 all use deep model.

Among different deep learning models, convolution neural networks (CNNs) were extensively used for the classification of small images like digits, characters [53]. This model starts to gain popularity after being used for the classification of large images in ImageNet challenge [54]. When CNN was first time considered for document image binarization [55], it was observed, a CNN with appropriate architecture can outperform deep MLP with no convolution. In a contemporary work [56], authors use CNN for semantic segmentation where, each pixel of the input image is classified in n different classes. For binarization the value of n will be 2. To serve their purpose, they transform convolutional networks to Fully convolution neural network (FCNN). However, this model did not perform well when evaluated using DIBCO datasets. Later Tensmeyer and Martinez [57], Vo et al. [58], and Peng et al. [59] also use FCNN to learn features from different scale. In contrast to [57], where authors developed single method, in [58], authors use three different networks that learn independently from input image at different scale. These networks are trained parallelly. Finally, the outputs of these models are combined to generate the final result. This model is found effective in both fine and coarse prediction. These FCNN based methods fail to ensure smoothness in output images. Therefore, require post processing. Considering this fact, in [60], authors use fully connected CRF for post processing whereas in [61], authors associate the smoothness term with the predicted probability of each pixel. This helps to learn the pixel probability along with the output smoothness and thus eliminates the need of post processing.

Encoder-decoder models are also used for DIB [62, 63]. However, the number of methods that use encoder-decoder models for DIB is less compares to CNN models. These models compress the input image and generate the code. Then decompress the code to reconstruct the image in the same resolution as the input. The method described in [63] uses an end-to-end selectional encoder-decoder architecture for pixel classification. In this work, authors use different blocks for their model and observe residual block performs well. They also identify the performance deference present between the models trained with specific document set and generalized set. This model suffers in presence of thin and weak strokes. In [62], authors train the encoder-decoder model using synthetic data. During training, authors consider additional layers at smaller resolution. Authors in [64] use the U-Net architecture at the global and local level to capture contextual information and local information respectively for pixel classification. As per the experimental results, this idea turns out to be useful compared to some other deep learning models on a few occasions.

Although these models perform better compared to the traditional methods, these require huge data for training. Therefore, insufficient training data may affect their testing performance. To address this issue authors in [65], use pretrained U-Net models to perform basic image processing tasks like erosion, dilation, canny edge detection, histogram equalization, otsu binarization, etc., and prepare the final mode for binarization by cascading these pretrained models. The authors report state-of-the-art result for DIBCO 2017 dataset. Besides that, Generative adversarial networks (GANs) are also used to generate realistic synthetic data for training. For example, in [66], authors use CycleGAN for synthetic data generation for better training which results in 1.4% increase in F-measure. In [67], authors use GAN discriminator to improve the efficiency of a binarization model. Authors in [68], introduce a clustering based method which ensures the labeling of divers data. This method decreases the amount of data required for training by 50% in cost of slight loss in accuracy.

Recurrent neural networks (RNNs) are also considered for document image binarization. However, these methods fail to achieve state-of-the-art result. One such method is reported in [69]. In this work, authors consider four bidirectional long short-term memory (BLSTM) networks to extract features from input images and before prediction, the extracted features are combined. Authors in [70] apply grid LSTMs and report significantly better performance.

Until the last few years, classical binarization methods [31, 71] produced state-of-the-art results. Recently, Deep learning based methods have stared gaining popularity due to their outperforming performances. In DIBCO 2017 [13] competition, these methods secure all the top positions. However, the top two methods in HDIBCO 2018 and DIBCO 2019 did not consider deep models. Although deep learning based methods showcased better performance, these are slow and require costly hardware (like Graphical Processing Unit) support to achieve reasonable speed [72]. This is the reason for that implementation of many binarization based applications became difficult. For example, binarization on mobile devices. Deep methods also suffer due to the scarcity of labeled training data. Although some initial attempts to address this issue has been made [66], there still exist a large gap to address.

2.5 Evaluation Techniques

Evaluation of such methods requires database of image with ground truth images and evaluation metrics. In this section, first description of some standard databases is presented then some popular metrices used for performance assessment of binarization techniques are discussed.

2.5.1 Standard Databases for DIB

In literature, a good number of standard databases are made available for the evaluation of different binarization techniques. Among these, databases that are made available through DIBCO competitions [14, 73] are very popular. Since 2009, these competitions are arranged [16] and each competition comes with a new set of data with different types of degradations. Most of these datasets contain around 10–12 printed or handwritten documents collected from different libraries. Almost all the degraded document image binarization methods consider these

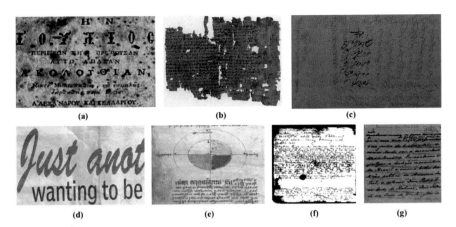

Fig. 2.4 Presents some sample images from (**a, b**) DIBCO 2019 dataset, (**c**) PHIBD dataset, (**d**) CMATERdb 6 dataset, (**e**) ISOS datset, (**f**) Bickley Diary dataset, and (**g**) time-quality binarization competition dataset

databases for evaluating their performance. Besides DIBCO datasets, few other datasets are there which gain less popularity. These include, The Persian Heritage Image Binarization Dataset (PHIBD) [74, 75] which contain 15 handwritten document images with several degradations written in Persian language, CMATERdb6 [76], which consist of 5 color images written in different languages, the Bickley diary dataset [71] consists of 7 page images with several degradations taken from a diary of 1920. However, Bickley diary dataset has started to gain attention in recent days [58, 77]. Recently, in the Time-Quality Binarization Competition [72] three new datasets are used to evaluate the participating methods, which were made available for download. This dataset contains total 104 degraded page images.

Another dataset containing images with mostly bleed through type of degradation is made available with the name Irish Script on Screen (ISOS) bleed-through dataset [78]. In this dataset, the pixels in ground truth images are labeled as foreground, background, and bleed through. Although the objective of this dataset is to train models to deal with bleed through degradation, but this dataset can also be used for the performance evaluation of the binarization methods in presence of bleed through. Figure 2.4 presents sample images taken from these datasets.

In addition to these paper document images, a completely divers category of documents called Palm Leaf Manuscripts (PLMs) are also used for performance evaluation of binarization methods. These documents also suffer from degradations like fade ink, bleed through, low contrast, stains, dark background, etc. (see Fig. 2.5). In a work [79], authors repot that recent binarization methods fail to obtain state-of-the-art result whereas few classical methods obtain comparably better result. One such dataset is the AMADI_Lontar Set [80] which contains 100 Balinese palm leaf manuscript images. In ICHFR 2018 competition [81], a similar dataset is used which contains 61 Sudanese and 46 Khmer PLMs.

(a)

(b)

Fig. 2.5 Presents some sample PLMs from (**a**) AMADI_LontarSet, and (**b**) ICHFR 2018

2.5.2 *Performance Metrics*

There are many approaches to assess the performance of a binarization method [82]. Among these the approach of measuring the closeness of the generated output to the GT image is popular. The metrices commonly used is this approach are as follows,

F-Measure (FM)

FM [14] is the harmonic mean of recall and precision. This can be expressed as

$$FM = 2 \times Recall \times Precision \big/ Recall + Precision \qquad (2.13)$$

Here, *Precision = True positive/(True positive + False positive)*, and *Recall = True positive/(True positive + False negative)*. The count of *True positive*, *False positive*, and *False negative* cases for an output $n \times m$ binary image I_b generated by an arbitrary binarization method can be computed with respect to the corresponding GT image I_{gt} as $True\ positive = \sum\limits_{x=1}^{n} \sum\limits_{y=1}^{m} I_b(x,y) \wedge I_{gt}(x,y)$, $False\ positive = \sum\limits_{x=1}^{n} \sum\limits_{y=1}^{m} I_b(x,y) \wedge \sim I_{gt}(x,y)$, and $False\ negative = \sum\limits_{x=1}^{n} \times \sum\limits_{y=1}^{m} \sim I_b(x,y) \wedge I_{gt}(x,y)$. This metric helps to penalize when the binarization method fails to maintain the proportion between *False positive* and *False negative*.

Peak Signal-to-Noise Ratio (PSNR)

PSNR [83] refers to the ratio between the maximum possible power of the signal and the power of noise that affects the quality of the signal. This metric indicates the efficiency of the binarization based on the visual quality of the generated output [84]. This can be represented as,

$$PSNR = 10\,log_{10}\left(\frac{1}{e^2}\right) \tag{2.14}$$

Here, $e^2 = \frac{1}{nm}\sum\limits_{x=1}^{n}\sum\limits_{y=1}^{m}\left((I_b(x,y) - I_{gt}(x,y))^2\right)$. It measures how close the generated output I_b is to the GT image I_{gt}. The increment in the value of PSNR indicates closeness of these two images. Additionally, increase of pixelwise accuracy also results in monotonous increase of PSNR as it considers total percentage of error e^2.

Distance Reciprocal Distortion (DRD)

This metric was introduced in DIBCO 2011 competition [85]. The performance assessment of a binarization method generated by this metric is very close to how bad the binarization error is from human perception [86]. This can be expressed as,

$$DRD = \frac{1}{S_G}\sum_{x}\sum_{y}DRD_{xy}\left|I_b(x,y) - I_{gt}(x,y)\right| \tag{2.15}$$

$$DRD_{xy} = \sum_{r=-2}^{2}\sum_{c=-2}^{2}W_{rc}\left|I_b(x+r,y+c) - I_{gt}(x+r,y+c)\right| \tag{2.16}$$

where, S_G is the number of 8×8 non-uniform patches of I_{gt}, $\sum\limits_{rc}W_{rc} = 1$. Lower DRD value indicates good result. DRD assigns more penalty to a output image where pixel errors appear in cluster or in close distance whereas it assigns lesser penalty when the errors appear in scattered way even the percentage of error is same. This is just like a cluster of wrong pixels attracts the attention of human eye more than an isolated one.

Pseudo-F-Measure (PseudoFM)

PseudoFM [87] is the harmonic mean of pseudo-precision (pp), and pseudo-recall (pr). The difference between {precision, recall} and {pseudo-precision, pseudo-recall}is that in pseudo-precision, and pseudo-recall every pixel is assigned a weight based on its position. This spatial position based weight for a pixel is computed with the help of the GT image. This metric tells how bad an error is in a particular location. PseudoFM can be expressed as follows,

$$\text{PseudoFM} = 2 \times pp \times pr/_{pp+pr} \tag{2.17}$$

$$pp = \frac{\sum\limits_{(x,y)} I_b(x,y) I_{gt}(x,y) WH_p(x,y)}{I_b(x,y) WH_p(x,y)} \quad pr = \frac{\sum\limits_{(x,y)} I_b(x,y) I_{gt}(x,y) WH_r(x,y)}{I_{gt}(x,y) WH_r(x,y)} \tag{2.18}$$

Here, WH_p penalizes the false positives more which appear near the foreground region or within the foreground components. This is because noise near the foreground region disturb the readability of the foreground text. Noise appear within foreground components results in attaching foreground texts. WH_r assign high penalty to the false negatives that appear on the skeleton of the foreground stroke as these cause discontinuity. Therefore, binarization results with discontinuity present in the foreground stroke incur more error than one with thinner but connecting strokes. Here, the false positive errors are considered to be of four types; text enlargement, touching text, background noise, and false detection, whereas the false negative errors are grouped into three categories; broken text, missed text, and partially missed text.

There will always be a debate on the proper way to assess different binarization methods, given that researchers always do not agree to follow GT comparison based approach [82]. However, the majority of published work has followed this approach to assess performance using GT images and metrics provided by DIBCO. In [72], running time is included to compare the binarization method for the first time.

References

1. Likforman-Sulem, L., Zahour, A., Taconet, B.: Text line segmentation of historical documents: a survey. Int. J. Doc. Anal. Recognit. **9**, 123–138 (2007)
2. van Kempen, G.M.P., van Vliet, L.J., Verveer, P.J., van der Voort, H.T.M.: A quantitative comparison of image restoration methods for confocal microscopy. J. Microsc. **185**(3), 354–365 (1997)
3. Sulaiman, A., Omar, K., Nasrudin, M.F.: A database for degraded Arabic historical manuscripts. In: 2017 6th International Conference on Electrical Engineering and Informatics (ICEEI), pp. 1–6 (2017)
4. Mustafa, W.A., Yazid, H.: Illumination and contrast correction strategy using bilateral filtering and binarization comparison. J. Telecommun. Electron. Comput. Eng. **8**(1), 67–73 (2016)
5. Sulaiman, A., Omar, K., Nasrudin, M.F.: Degraded historical document binarization: a review on issues, challenges, techniques, and future directions. J. Imaging. **5**(4), 48 (2019)
6. Dougherty, G.: Digital Image Processing for Medical Applications. Cambridge University Press (2009)
7. Verma, R., Ali, J.: A comparative study of various types of image noise and efficient noise removal techniques. Int. J. Adv. Res. Comput. Sci. Softw. Eng. **3**(10) (2013)
8. Astola, J., Kuosmanen, P.: Fundamentals of Nonlinear Digital Filtering. CRC Press (2020)
9. Dhananjay, K.T.: Digital Image Processing (Using MATLAB Codes) (Jul 2013)
10. Kamboj, P., Rani, V.: A brief study of various noise model and filtering techniques. J. Glob. Res. Comput. Sci. **4**(4), 166–171 (2013)
11. Singh, P., Shree, R.: A comparative study to noise models and image restoration techniques. Int. J. Comput. Appl. **149**(1), 18–27 (2016)

12. Farahmand, A., Sarrafzadeh, H., Shanbehzadeh, J.: Document Image Noises and Removal Methods (2013)
13. Pratikakis, I., Zagoris, K., Barlas, G., Gatos, B.: ICDAR2017 competition on document image binarization (DIBCO 2017). In: 2017 14th IAPR International Conference on Document Analysis and Recognition (ICDAR), vol. 1, pp. 1395–1403 (2017)
14. Pratikakis, I., Zagoris, K., Karagiannis, X., Tsochatzidis, L., Mondal, T., Marthot-Santaniello, I.: ICDAR 2019 competition on document image binarization (DIBCO 2019). In: International Conference on Document Analysis and Recognition (ICDAR), pp. 1547–1556 (2019)
15. Binmakhashen, G.M., Mahmoud, S.A.: Document layout analysis: a comprehensive survey. ACM Comput. Surv. 52(6), 1–36 (2019)
16. Gatos, B., Ntirogiannis, K., Pratikakis, I.: ICDAR 2009 document image binarization contest (DIBCO 2009). In: 10th International Conference on Document Analysis and Recognition, 2009. ICDAR'09, pp. 1375–1382 (2009)
17. Pratikakis, I., Gatos, B., Ntirogiannis, K.: ICDAR 2013 document image binarization contest (DIBCO 2013). In: 2013 12th International Conference on Document Analysis and Recognition (ICDAR), pp. 1471–1476 (2013)
18. Lins, R.D., de Almeida, M.M., Bernardino, R.B., Jesus, D., Oliveira, J.M.: Assessing binarization techniques for document images. In: Proceedings of the 2017 ACM Symposium on Document Engineering, pp. 183–192 (2017)
19. Otsu, N.: A threshold selection method from gray-level histograms. IEEE Trans. Syst. Man Cybern. 9(1), 62–66 (1979)
20. Kittler, J., Illingworth, J., Föglein, J.: Threshold selection based on a simple image statistic. Comput. Vision Graph. Image Process. 30(2), 125–147 (1985)
21. Niblack, W.: An Introduction to Digital Image Processing. Strandberg Publishing Company (1985)
22. Das, B., Bhowmik, S., Saha, A., Sarkar, R.: An adaptive foreground-background separation method for effective binarization of document images. In: International Conference on Soft Computing and Pattern Recognition, pp. 515–524 (2016)
23. Sauvola, J., Pietikäinen, M.: Adaptive document image binarization. Pattern Recogn. 33(2), 225–236 (2000)
24. Wolf, C., Jolion, J.-M., Chassaing, F.: Text localization, enhancement and binarization in multimedia documents. In: 2002 International Conference on Pattern Recognition, vol. 2, pp. 1037–1040 (2002)
25. Gatos, B., Pratikakis, I., Perantonis, S.J.: Adaptive degraded document image binarization. Pattern Recogn. 39(3), 317–327 (2006)
26. Trier, O.D., Jain, A.K.: Goal-directed evaluation of binarization methods. IEEE Trans. Pattern Anal. Mach. Intell. 17(12), 1191–1201 (1995)
27. Sezgin, M.: Survey over image thresholding techniques and quantitative performance evaluation. J. Electron. Imaging. 13(1), 146–168 (2004)
28. Mishra, A., Alahari, K., Jawahar, C.V.: Unsupervised refinement of color and stroke features for text binarization. Int. J. Doc. Anal. Recognit. 20, 1–17 (2017)
29. Chen, Y., Wang, L.: Broken and degraded document images Binarization. Neurocomputing. 237, 272–280 (2017)
30. Lu, D., Huang, X., Liu, C., Lin, X., Zhang, H., Yan, J.: Binarization of degraded document image based on contrast enhancement. In: Control Conference (CCC), 2016 35th Chinese, pp. 4894–4899 (2016)
31. Howe, N.R.: Document binarization with automatic parameter tuning. Int. J. Doc. Anal. Recognit. 16(3), 247–258 (2013)
32. Boykov, Y., Kolmogorov, V.: An experimental comparison of min-cut/max-flow algorithms for energy minimization in vision. IEEE Trans. Pattern Anal. Mach. Intell. 26(9), 1124–1137 (2004)
33. Kolmogorov, V., Zabin, R.: What energy functions can be minimized via graph cuts? IEEE Trans. Pattern Anal. Mach. Intell. 26(2), 147–159 (2004)

34. Ayyalasomayajula, K.R., Brun, A.: Document binarization using topological clustering guided Laplacian energy segmentation. In: 2014 14th International Conference on Frontiers in Handwriting Recognition, pp. 523–528 (2014)
35. Pratikakis, I., Gatos, B., Ntirogiannis, K.: ICFHR 2012 competition on handwritten document image binarization (H-DIBCO 2012). In: 2012 International Conference on Frontiers in Handwriting Recognition (ICFHR), pp. 817–822 (2012)
36. Ntirogiannis, K., Gatos, B., Pratikakis, I.: ICFHR2014 competition on handwritten document image binarization (H-DIBCO 2014). In: 2014 14th International Conference on Frontiers in Handwriting Recognition (ICFHR), pp. 809–813 (2014)
37. Pratikakis, I., Zagoris, K., Barlas, G., Gatos, B.: ICFHR2016 Handwritten Document Image Binarization Contest (H-DIBCO 2016). In: 2016 15th International Conference on Frontiers in Handwriting Recognition (ICFHR), pp. 619–623 (2016)
38. Peng, X., Setlur, S., Govindaraju, V., Sitaram, R.: Markov random field based binarization for hand-held devices captured document images. In: Proceedings of the 7th Indian Conference on Computer Vision, Graphics and Image Processing, pp. 71–76 (2010)
39. Su, B., Lu, S., Tan, C.L.: A learning framework for degraded document image binarization using Markov random field. In: Proceedings of the 21st International Conference on Pattern Recognition (ICPR2012), pp. 3200–3203 (2012)
40. Kuk, J.G., Cho, N.I.: Feature based binarization of document images degraded by uneven light condition. In: 10th International Conference on Document Analysis and Recognition, 2009. ICDAR'09, pp. 748–752 (2009)
41. Wolf, C., Doermann, D.: Binarization of low quality text using a Markov random field model. In: Proceedings of the 16th International Conference on Pattern Recognition, 2002, vol. 3, pp. 160–163 (2002)
42. Papamarkos, N.: A neuro-fuzzy technique for document binarisation. Neural Comput. Appl. **12**(3–4), 190–199 (2003)
43. Rabelo, J.C.B., Zanchettin, C., Mello, C.A.B., Bezerra, B.L.D.: A multi-layer perceptron approach to threshold documents with complex background. In: 2011 IEEE International Conference on Systems, Man, and Cybernetics, pp. 2523–2530 (2011)
44. Kefali, A., Sari, T., Bahi, H.: Foreground-background separation by feed-forward neural networks in old manuscripts. Informatica. **38**(4), 329 (2014)
45. Pastor-Pellicer, J., Zamora-Martínez, F., España-Boquera, S., Castro-Bleda, M.J.: F-measure as the error function to train neural networks. In: Advances in Computational Intelligence: 12th International Work-Conference on Artificial Neural Networks, IWANN 2013, Puerto de la Cruz, Tenerife, Spain, 12–14 Jun 2013, Proceedings, Part I, vol. 12, pp. 376–384 (2013)
46. Trier, O.D., Taxt, T.: Evaluation of binarization methods for document images. IEEE Trans. Pattern Anal. Mach. Intell. **17**(3), 312–315 (1995)
47. Badekas, E., Papamarkos, N.: Document binarisation using Kohonen SOM. IET Image Process. **1**(1), 67–84 (2007)
48. Badekas, E., Papamarkos, N.: Optimal combination of document binarization techniques using a self-organizing map neural network. Eng. Appl. Artif. Intell. **20**(1), 11–24 (2007)
49. Mitianoudis, N., Papamarkos, N.: Document image binarization using local features and Gaussian mixture modeling. Image Vis. Comput. **38**, 33–51 (2015)
50. Jana, P., Ghosh, S., Bera, S.K., Sarkar, R.: Handwritten document image binarization: an adaptive K-means based approach. In: 2017 IEEE Calcutta Conference (CALCON), pp. 226–230 (2017)
51. Jana, P., Ghosh, S., Sarkar, R., Nasipuri, M.: A fuzzy C-means based approach towards efficient document image binarization. In: 2017 Ninth International Conference on Advances in Pattern Recognition (ICAPR), pp. 1–6 (2017)
52. Bhowmik, S., Sarkar, R., Das, B., Doermann, D.: GiB: A game theory inspired binarization technique for degraded document images. IEEE Trans. Image Process. **28**(3) (2019). https://doi.org/10.1109/TIP.2018.2878959

53. Simard, P.Y., Steinkraus, D., Platt, J.C.: Best practices for convolutional neural networks applied to visual document analysis. In: ICDAR, 2003, vol. 3, pp. 958–962 (2003)
54. Russakovsky, O., et al.: Imagenet large scale visual recognition challenge. Int. J. Comput. Vis. **115**, 211–252 (2015)
55. Pastor-Pellicer, J., España-Boquera, S., Zamora-Martínez, F., Afzal, M.Z., Castro-Bleda, M.J.: Insights on the use of convolutional neural networks for document image binarization. In: Advances in Computational Intelligence: 13th International Work-Conference on Artificial Neural Networks, IWANN 2015, Palma de Mallorca, Spain, 10–12 Jun 2015. Proceedings, Part II, vol. 13, pp. 115–126 (2015)
56. Long, J., Shelhamer, E., Darrell, T.: Fully convolutional networks for semantic segmentation. In: Proceedings of the IEEE Conference on Computer Vision and Pattern Recognition (2015) pp. 3431–3440
57. Tensmeyer, C., Martinez, T.: Document Image Binarization with Fully Convolutional Neural Networks. arXiv Prepr. arXiv1708.03276 (2017)
58. Vo, Q.N., Kim, S.H., Yang, H.J., Lee, G.: Binarization of degraded document images based on hierarchical deep supervised network. Pattern Recogn. **74**, 568–586 (2018)
59. Peng, X., Wang, C., Cao, H.: Document binarization via multi-resolutional attention model with DRD loss. In: 2019 International Conference on Document Analysis and Recognition (ICDAR), pp. 45–50 (2019)
60. Krähenbühl, P., Koltun, V.: Efficient inference in fully connected crfs with Gaussian edge potentials. In: Adv. Neural Inf. Process. Syst., vol. 24 (2011)
61. Ayyalasomayajula, K.R., Malmberg, F., Brun, A.: PDNet: semantic segmentation integrated with a primal-dual network for document binarization. Pattern Recognit. Lett. **121**, 52–60 (2018)
62. Peng, X., Cao, H., Natarajan, P.: Using convolutional encoder-decoder for document image binarization. In: 2017 14th IAPR International Conference on Document Analysis and Recognition (ICDAR), pp. 708–713 (2017)
63. Calvo-Zaragoza, J., Gallego, A.-J.: A selectional auto-encoder approach for document image binarization. Pattern Recogn. **86**, 37–47 (2019)
64. Huang, X., Li, L., Liu, R., Xu, C., Ye, M.: Binarization of degraded document images with global-local U-Nets. Optik (Stuttg). **203**, 164025 (2020)
65. Kang, S., Iwana, B.K., Uchida, S.: Cascading modular u-nets for document image binarization. In: 2019 International Conference on Document Analysis and Recognition (ICDAR), pp. 675–680 (2019)
66. Tensmeyer, C., Brodie, M., Saunders, D., Martinez, T.: Generating realistic binarization data with generative adversarial networks. In: 2019 International Conference on Document Analysis and Recognition (ICDAR), pp. 172–177 (2019)
67. Zhao, J., Shi, C., Jia, F., Wang, Y., Xiao, B.: Document image binarization with cascaded generators of conditional generative adversarial networks. Pattern Recogn. **96**, 106968 (2019)
68. Krantz, A., Westphal, F.: Cluster-based sample selection for document image binarization. In: 2019 International Conference on Document Analysis and Recognition Workshops (ICDARW), vol. 5, pp. 47–52 (2019)
69. Afzal, M.Z., Pastor-Pellicer, J., Shafait, F., Breuel, T.M., Dengel, A., Liwicki, M.: Document image binarization using lstm: a sequence learning approach. In: Proceedings of the 3rd International Workshop on Historical Document Imaging and Processing, pp. 79–84 (2015)
70. Westphal, F., Lavesson, N., Grahn, H.: Document image binarization using recurrent neural networks. In: 2018 13th IAPR International Workshop on Document Analysis Systems (DAS), pp. 263–268 (2018)
71. Su, B., Lu, S., Tan, C.L.: Robust document image binarization technique for degraded document images. IEEE Trans. Image Process. **22**(4), 1408–1417 (2013)
72. Lins, R.D., Bernardino, R., Jesus, D.M.: A quality and time assessment of binarization algorithms. In: 2019 International Conference on Document Analysis and Recognition (ICDAR), pp. 1444–1450 (2019)

73. ICDAR. DIBCO competition datasets. https://vc.ee.duth.gr/dibco2019/ (2019). Accessed 27 Feb 2023
74. Ayatollahi, S.M., Nafchi, H.Z.: Persian heritage image binarization competition (PHIBC 2012). In: 2013 First Iranian Conference on Pattern Recognition and Image Analysis (PRIA), pp. 1–4 (2013)
75. Hossein Ziaie Nafchi, M.C., Ayatollahi, S.M., Moghaddam, R.F.: PHIBD 2012. http://www. iapr-tc11.org/mediawiki/index.php/Persian_Heritage_Image_Binarization_Dataset_ (PHIBD_2012)
76. Mollah, A.F., Basu, S., Nasipuri, M.: Computationally efficient implementation of convolution-based locally adaptive binarization techniques. In: Wireless Networks and Computational Intelligence: 6th International Conference on Information Processing, ICIP 2012, Bangalore, India, 10–12 Aug 2012. Proceedings, pp. 159–168 (2012)
77. Nafchi, H.Z., Moghaddam, R.F., Cheriet, M.: Phase-based binarization of ancient document images: model and applications. IEEE Trans. Image Process. 23(7), 2916–2930 (2014)
78. Rowley-Brooke, R., Pitié, F., Kokaram, A.: A ground truth bleed-through document image database. In: Proceedings of the Second International Conference on Theory and Practice of Digital Libraries (TPDL 2012), Paphos, Cyprus, 23–27 Sept 2012, vol. 2, pp. 185–196 (2012)
79. Kesiman, M.W.A., Prum, S., Burie, J.-C., Ogier, J.-M.: An initial study on the construction of ground truth binarized images of ancient palm leaf manuscripts. In: 2015 13th International Conference on Document Analysis and Recognition (ICDAR), pp. 656–660 (2015)
80. Kesiman, M.W.A., Burie, J.-C., Wibawantara, G.N.M.A., Sunarya, I.M.G., Ogier, J.-M.: AMADI_LontarSet: the first handwritten Balinese palm leaf manuscripts dataset. In: 2016 15th International Conference on Frontiers in Handwriting Recognition (ICFHR), pp. 168–173 (2016)
81. Kesiman, M.W.A., et al.: ICFHR 2018 competition on document image analysis tasks for southeast Asian palm leaf manuscripts. In: 2018 16th International Conference on Frontiers in Handwriting Recognition (ICFHR), pp. 483–488 (2018)
82. Tensmeyer, C., Martinez, T.: Historical document image binarization: a review. SN Comput. Sci. 1(3), 173 (2020)
83. Bera, S.K., Ghosh, S., Bhowmik, S., Sarkar, R., Nasipuri, M.: A Non-parametric Binarization Method Based on Ensemble of Clustering Algorithms. Multimed. Tools Appl. 80, 7653–7673 (2020). https://doi.org/10.1007/s11042-020-09836-z
84. Yu, J., Bhanu, B.: Evolutionary feature synthesis for facial expression recognition. Pattern Recogn. Lett. 27(11), 1289–1298 (2006). https://doi.org/10.1016/j.patrec.2005.07.026
85. Pratikakis, I,, Gatos, B., Ntirogiannis, K.: ICDAR 2011 Document Image Binarization Contest (DIBCO 2011). In: ICDAR, Sept 2011, pp. 1506–1510 (2011)
86. Lu, H., Kot, A.C., Shi, Y.Q.: Distance-reciprocal distortion measure for binary document images. IEEE Signal Process. Lett. 11(2), 228–231 (2004)
87. Ntirogiannis, K., Gatos, B., Pratikakis, I.: Performance evaluation methodology for historical document image binarization. IEEE Trans. Image Process. 22(2), 595–609 (2012)

Chapter 3
Document Region Segmentation

Abstract Documents possess a hierarchy of physical modules which is referred as the structure of the document. Document structure includes both geometrical and logical aspects. Geometrical aspect of a document structure is the geometrical representation of various constituent regions' boundaries and their spatial arrangements. The logical one refers to the labelling of these regions. For analysing the layout of an input document image, it is essential to extract the geometrical structure of the document first. Therefore, in this chapter, various document region segmentation methods are discussed along with their pros and cons. Additionally, this chapter discusses various standard datasets and different evaluation metrices developed for the evaluation of the said methods.

Keywords Document region segmentation · Pixel analysis · Connected component analysis · Local region analysis · CNN · RCNN · Deep learning

As said earlier, any document image possesses two types of structure, first the geometrical or physical structure and semantic or logical structure. Geometrical structure of a document refers to the spatial arrangement of different regions. It includes, the boundary information of the regions and their spatial relationship. Semantic or logical structure of a document refers to the functional labelling of the regions. In this section, different methods developed for document region segmentation are discussed.

3.1 Different Document Region Segmentation Methods

Research on layout analysis was started long back, and since then several methods are introduced by the researchers for that purpose. However, the continuous advancement of multimedia technology causes the generation of more complex and decorative paper documents. As a result, documents are becoming more expressive to the readers and able to attract their attention as well. But this, in turn, limits the applicability of many popular and age-old DLA techniques, introduced earlier in this domain [1, 2], to these modern paper documents. Considering this fact, recently, Kise in [3], classify the segmentation techniques according to the types of layout (i.e., overlapping and non-overlapping) these can handle. On the other hand, in [4], Eskenazi et al. present a very comprehensive discussion on different DLA techniques and classify them, according to the reason of their limitations, into three classes namely, *algorithm constrained, parameter constrained*, and *potentially unconstrained*. Beside these contemporary classifications, a very general and widely accepted classification of these methods is *top-down, bottom-up*, and *hybrid* [5]. This classification is based on how these methods segment a particular document image. For example, *top-down* methods start with the entire image and split it into regions, and then split the regions into text lines and the text lines into words. In contrast, *bottom-up* methods start at pixel level or component level, then keep merging them to form words, and words to text lines, and text lines to regions. *Hybrid* methods consider both of these concepts. A detailed discussion on this classification is presented in [6, 7].

However, it is hard to make a clear boundary to group these methods into some specific categories. Here, for better understanding, these methods are grouped based on the exact elements of the input documents these methods consider to start the analysis for segmentation. Accordingly, four groups are prepared: *Pixel analysis-based methods, Connected component analysis-based methods, Local region analysis-based methods*, and *Hybrid methods*.

3.1.1 Pixel Analysis-Based Methods

The methods from this category involve pixel-level operations at the initial phase for the segmentation of the input document images. Based on the type of pixel they consider, these methods can further be classified as (a) *foreground pixel analysis-based methods*, (b) *background pixel analysis-based methods*, and (c) *foreground-background pixel analysis-based methods*.

Foreground pixel analysis-based methods try to combine the foreground or data pixels by filling up the gap in between them so that an input image can be segmented into various regions. The earliest proposal of this category of methods is the *run-length smearing algorithm* (RLSA) [8] also known as run-length smoothing. After that many page segmentation techniques [9, 10] are introduced following this method. Any RLSA based method tries to merge the foreground pixels by filling the inter-component gaps. Here, the run-length smearing is applied both horizontally and vertically to generate the final result. But these methods work well mostly for Manhattan layout, thus turning out to be inappropriate in the latter days when documents with non-Manhattan layouts become a trend. To overcome this issue, in [11] Sun et al. introduce a modified version of it called *constraint run-length algorithm* (CRLA), which is capable of dealing with documents having a non-Manhattan layout. However, smearing algorithms suffer badly when the input document images are skewed. There is another group of methods that can be placed in this category, called *mathematical morphology-based methods*. These methods can be considered as a more generalized version of the smearing algorithms as they try to merge the foreground pixels by performing various morphological operations to generate the regions. The fundamental operations performed by these methods are morphological erosion, i.e. $X \ominus Y$ and dilation, i.e., $X \oplus Y$. Based on these two more compound operations are also commonly used i.e. morphological opening (see Eq. 3.1) and morphological closing (see Eq. 3.2).

$$X \circ Y = (X \ominus Y) \oplus Y \tag{3.1}$$

$$X \cdot Y = (X \oplus Y) \ominus Y \tag{3.2}$$

The opening eliminates the foreground objects which are smaller with respect to the structuring element. Whereas the closing fills the gaps. Morphology-based methods generally use either one or both as per their preferred order. For example, Chen et al. in [12], apply the closing operation on the binarized image to generate the initial regions and then classify these as text or non-text based on their size. Bloomberg in [13], introduces a multi-resolution morphology-based method to eliminate the texts iteratively so that at the end of the process, the resultant image is left with the half-tone components. Later Bukhari et al. in [14], propose an improved version of this algorithm. However, the performance of morphology-based methods heavily depends on the dimension of the structuring element.

Background pixel analysis methods try to use the background pixel streams or white streams so that the region delimiters are identified and the input image can be segmented. Projection profile-based methods [15, 16] are the earliest ones belonging to this category. These methods compute the vertical and horizontal projection profiles to find out the white delimiters. Similar to the RLSA based methods, these are only applicable to the document images with Manhattan layout and suffer when the input images are skewed [17]. In [18], Normand and Viard-Gaudin present an extended RLSA method that can deal with skew, and in [19, 20] Kise et al. propose two background pixel analysis methods that are applicable to non-Manhattan layouts. In [19], authors represent the white streams using thin lines as thinning of the background preserves the connectivity. Thereafter they identify the closed loops that enclose different regions. However, in [20], the approximated area in the Voronoi diagram is used to estimate background white stream. Both of these methods can also deal with skewed images.

Foreground-background pixel analysis methods [21] consider both the foreground and background pixels to generate the regions. For example, Antonacopoulos and Ritchings in [22], initially perform foreground pixel smearing then calculate the background pixel streams so that the region separator or region boundaries can be identified. Pavlidis and Zhou [23] perform horizontal smearing of the foreground pixels to get the column delimiters. However, most of these methods require prior knowledge about the input documents like font size, gap size of texts, etc. Mehri et al. in [24] first generates the background and foreground superpixels using a simple linear iterative clustering technique and then classify those superpixels based on Gabor-based texture features using Support vector machine (SVM) to produce the segmentation result. Recently, Chen et al., in [25], use a convolutional auto-encoder to extract features from pixel intensity and then apply these features to train a support vector machine (SVM) classifier so that it can be used for classifying the pixels of the input images either as periphery, background, text block, or decoration. Although this method is skew-invariant and has no effect on the font size variation, this method is highly parametric and time-consuming. Moreover, the authors have not discussed the effect of different parameters on the overall performance.

Recently, deep learning based methods become dominant in all most every field. In document image segmentation also, deep models showcase their effectiveness. Most of the deep models perform segmentation by performing classification at pixel level. Among all the deep methods, FCNNs are mostly preferred by the researcher. One important reason for that reported in [26] is, these models are relatively light weight. Authors in [27] design a unified multimodal FCNN which simultaneously detect the appearance based classes and semantic classes at pixel level. This network consists of an encoder to learn feature representation hierarchy, decoder to generate segmentation mask, auxiliary decoder that is considered only during training to image reconstruction, and the bridge connection is used for textual and visual representation. This model support both supervised and unsupervised training. Another work reported in [28], where, a CNN with encoder and decoder is used to segment historical documents. The encoder and decoder used in this model is completely developed based on convolution and deconvolution layers. Authors in [29], device a multitasking, open source, CNN architecture for historical document which can simultaneously performs baseline extraction, page extraction, layout analysis, figure and illustration extraction. In contrast to these complex models, a relatively simple CNN model with one convolution layer is proposed in [30] for pixel classification. The model obtains comparable result. Another very complex model is reported in [31], where, a hybrid spatial-channel attention model (HSCA-Net) is developed. This model consists of spatial attention, and channel attention modules, and lateral attention connections. The spatial attention module emphasises on the inter-region relationship to estimate the global context. The channel attention module automatically adjusts the channel feature responses depending on the contribution of the features from each channel. The lateral attention connection is used to include spatial and channel attention module in multiscale feature pyramid network and to retain the original features.

3.1.2 Connected Component Analysis-Based Methods

These methods analyze and combine the CCs to form the regions. For example, in [32], Fletcher et al. apply Hough transform to combine the components to form character strings. Gorman, in [2], uses inter-component distances to form the text blocks. For that, they compute K-nearest neighbor (KNN) for each component and use a distance threshold to combine them. Zlatopolsky [33] initially combines the components into text lines and estimates the horizontal and vertical distances to merge them into blocks. Similar to these methods, the concept of inter-component distances or components closeness is employed in [34, 35] to form the regions. Gatos et al. in [14], and Antonacopoulos et al. in [62], evaluate the performances of many of such methods for the segmentation of newspapers and other documents. Some recent methods of this category are briefed here. In [36] Zagoris et al. initially extract the CCs to obtain the foreground blocks by combining them and then classify each such block either as text or non-text using some textural features. Tran et al. recently propose a two-stage method consisting of a heuristic filter in [37] to classify the extracted CCs as text and non-text. Tran et al. in [38], follow an iterative classification approach to separate the text from non-text CCs. This method assumes non-text are the CCs that cause heterogeneity in the CCs present in a document image and thereby keep eliminating all the CCs which are heterogeneous to their neighbors in each iteration until a homogeneous region with only text CCs is found. They further classify the non-texts as noises, separators, and figures based on their structural information. Le et al., in [39], use the stroke-width, size, shape, and positional information of the CCs to classify the components using Adaboosting decision tree. CC analysis-based methods can handle the skewness of the documents if the inter-line gap is smaller than the inter-paragraph gap which may not be the case for all the documents [17].

3.1.3 Local Region Analysis-Based Methods

Methods of this category follow a two-step approach. Of these, the first step is used to extract features from different local regions whereas the second step classifies them using a supervised or unsupervised classification algorithm. For example, Oyedotun and Khashman, in [40, 41] partition the input image into some non-overlapping small regions and classify them based on some first-order texture features like median and modal pixel values along with GLCM based second-order texture features using a feed-forward neural network-based classifier. Lin et al. in [41] also follow the similar approach of partition first and classify next using GLCM based features. Journet et al. in [42], after partitioning the input image, estimate the orientation distribution for each local region to identify it as text or graphic. Acharyya and Kundu, in [43], apply an M-band wavelet to extract features and then use an unsupervised learning technique to cluster these into some desired groups. Although in many cases these techniques provide promising results, the performance of these techniques is highly sensitive to the process of local region extraction. Moreover, wavelet-based methods are computationally complex and find difficulties for the documents with fewer elements or where the size of the texts is similar to the size of non-texts [44].

Recently, authors have started to use object detection models to isolate local regions in document images. For that purpose, Fast-Region based CNN(F-RCNN) and Mask-RCNN(M-RCNN) are commonly used. F-RCNN initially, generates candidate regions using Region proposal network (RPN) and then classify the generated regions. M-RCNN additionally generates binary mask for each candidate region. For example, in [45], authors consider both F-RCNN and M-RCNN to detect document regions. Another method reported in [46], where authors apply only F-RCNN for region detection.

3.1.4   Hybrid Methods

This category of methods performs both pixel and component-level analysis to segment an input document image. Many of the recently proposed techniques follow this approach [17, 47, 48]. For example, Tran et al. in [44], initially do CC analysis to perform text non-text separation and then use adaptive mathematical morphology to segment the text regions. For text segmentation, they estimate the vertical and horizontal whitespaces. Their method has won the RDCL 2015 page segmentation competition [47] and in [59] they introduce a modified version of it, which has also won the RDCL 2017 page segmentation competition [10]. The method, called *Fraunhofer Newspaper Segmentation System,* winner of ICDAR 2009 page segmentation competition [49], also follows this approach. Recently, Vasilopoulos et al. in [17], perform a morphological region merging operation at foreground and background to identify the separators. They also perform a contour tracing algorithm to classify the CC. In this work, the authors have made many assumptions regarding the inter-component gaps like there will always be a gap between the components of different classes, the inter-character, and inter-line gaps will be uniform, etc. In [50], Lu et al. propose a probabilistic framework for document image segmentation. In this work, initially, the extracted CCs are filtered to separate the noise and obvious non-texts. From the remaining CCs, a probabilistic local text homogeneity (PLTH) map is constructed. Further, the map is thresholded to separate the text and non-text CCs. From the text CCs, the paragraphs are obtained by using a conventional bottom-up approach, and to obtain the non-text structures mathematical morphology is used. This method has won RDCL 2019 page segmentation competition [15]. Although most of these methods are not skew-invariant, for non-skewed documents, these are turned out to be the most efficient methods, even for document images with a non-Manhattan layout. Several such methods are examined by Antonacopoulos et al. in [47–49, 51].

There are some deep model based works recently reported which follow hybrid approach. For example, in [52], authors use four base line models which include; BERT a transformer based pre-trained model which was trained on a large text corpus, RoBERTa an extended version of BERT, LayoutLM is also a pre-trained language model which models both text and the layout, and finally F-RCNN. In this work, the first three models are applied at the token level and F-RCNN is used at the region level.

3.2 Available Dataset for Page Segmentation

In literature, many databases are made available for the evaluation of different DLA methods. For example, the *Newspaper and magazine images segmentation database* of the *UCI machine learning repository* [53] contains 101 scanned images taken from different Russian newspapers and magazines. This database contains pixel-level ground-truth of the scanned images indicating the text, background, and picture regions. Another database is prepared by the University of Washington (UW-III) [39, 54], which consists of 1600 images of English technical articles with region-level ground truth indicating text, math, and figure regions. The *MediaTeam Document Database* [36, 55] contains 512 scanned pages taken from different types of English documents with the region-level ground-truth indicating both the physical and logical structures of these document images.

PRImA Research Lab, University of Salford, UK has made available the most popular databases of this domain [56]. The first database contains 305 page images taken from contemporary technical articles and magazines. These images are used in ICDAR 2001, 2003, 2005, 2007, 2009, 2011, 2015, 2017, and 2019 page layout competitions for evaluation. The second database contains 70,000 historical page images of documents printed in seventeenth to twentieth centuries. This database is prepared under the project named *Improving access to text* (IMPACT). The pages in this database are used in ICDAR 2013 and 2015 historical page layout competitions. The ground truths for these page layout competitions are prepared in *Page analysis and ground truth elements* (PAGE) format [57].

Beside these databases, another database called BCE-Arabic dataset [58] is available for the said task which was jointly developed by Boston University, Cairo University, and Electronics Research Institute. This database contains 1833 page images taken from 180 books with ground-truths for evaluating different tasks like *text analysis, text vs. images, text vs. graphic elements, text vs. tables, single or double column text vs. images,* etc.

In addition to these databases, few more datasets are there which are not publicly available but used for evaluation purposes [50, 59]. For example, UNLV database (University of Nevada, Las Vegas, USA) [60] contains 2000 scanned images of magazines, reports, and newspapers. CNU-Korean and CNU-English databases [61] contain 50 Korean language color document images and 45 English color document images respectively.

Recently, a dataset named with PubLayNet [45] is made available specially to train and evaluate deep model based DLA methods. This dataset contains 3,60,000 document images with annotation. A summary of these databases is given in Table 3.1.

Table 3.1 List of some standard databases for document layout analysis

Ref.	Database	Type	Size	Language	Ground-truth
[53]	Newspaper and magazine images segmentation database-UCI machine learning repository	Contemporary	101 images	Russian	Yes
[54]	University of Washington (UW-III) database	Contemporary	1600 images	English	Yes
[55]	*MediaTeam Document Database*	Contemporary	512 images	English	Yes
[62]	PRImA database	Contemporary	1240 images	English	Yes
[56]	PRImA database	Historical	70,000 images	Multiple	Yes
[58]	BCE-Arabic dataset	Contemporary	1833 images	Arabic	Yes
[60]	UNLV database	Contemporary	2000 images	English	–
[60]	CNU database	Contemporary	95 images	Multiple	–
[45]	PubLayNet	Contemporary	360k images	English	–

3.3 Evaluation Metrices

The document image processing community has been providing a common platform for the newly developed layout analysis methods to verify their progress in addressing the utmost challenges in DLA by arranging different competitions. For example, the PRImA Research Lab, University of Salford, UK has been conducting such competitions in association with International Conference on Document Analysis and Recognition (ICDAR) since 2001 [49].

In addition to some well-known databases, these competitions also use an advanced framework for evaluating the performance of different layout analysis methods [51]. This framework maps the output regions obtained by a method with the corresponding ground truth regions to assess the performance considering the following errors,

- *Merge*: When for two separate ground truth regions, a single region is found in the output.
- *Split*: When for a single ground truth region, two separate regions are found in the output.
- *Miss*: When for any ground truth region, no corresponding region is found in the output.

- *Partial miss*: When the ground truth region is not completely overlapped with the corresponding output region.
- *False detection*: When for an output region, no ground truth region is present.

In addition to these segmentation errors, the miss-classification error is computed as,

- *Miss-classification*: When the type of ground truth region is different from the corresponding output region.

All these errors are quantified and then multiplied by the area of the affected region and different weights. These weights are assigned based on two levels of error significance: first one is the *context-dependent* significance and second one is the *scenarios*. For each scenario, there is an evaluation profile that contains different distributions of these weights [62].

References

1. Lee, S.-W., Ryu, D.-S.: Parameter-free geometric document layout analysis. IEEE Trans. Pattern Anal. Mach. Intell. **23**(11), 1240–1256 (2001)
2. O'Gorman, L.: The document spectrum for page layout analysis. IEEE Trans. Pattern Anal. Mach. Intell. **15**(11), 1162–1173 (1993)
3. Kise, K.: Page segmentation techniques in document analysis. In: Handbook of Document Image Processing and Recognition, pp. 135–175. Springer (2014)
4. Eskenazi, S., Gomez-Krämer, P., Ogier, J.-M.: A comprehensive survey of mostly textual document segmentation algorithms since 2008. Pattern Recogn. **64**, 1–14 (2017)
5. Binmakhashen, G.M., Mahmoud, S.A.: Document layout analysis: a comprehensive survey. ACM Comput. Surv. **52**(6), 1–36 (2019)
6. Okun, O., Dœrmann, D., Pietikainen, M.: Page Segmentation and Zone Classification: the State of the Art. DTIC Document (1999)
7. Bhowmik, S., Sarkar, R., Nasipuri, M., Doermann, D.: Text and non-text separation in offline document images: a survey. Int. J. Doc. Anal. Recognit. **21**(1–2), 1–20 (2018)
8. Wong, K.Y., Casey, R.G., Wahl, F.M.: Document analysis system. IBM J. Res. Dev. **26**(6), 647–656 (1982)
9. Tsujimoto, S., Asada, H.: Major components of a complete text reading system. Proc. IEEE. **80**(7), 1133–1149 (1992)
10. Fan, K.-C., Liu, C.-H., Wang, Y.-K.: Segmentation and classification of mixed text/graphics/ image documents. Pattern Recogn. Lett. **15**(12), 1201–1209 (1994)
11. Sun, H.-M.: Page segmentation for Manhattan and non-Manhattan layout documents via selective CRLA. In: Eighth International Conference on Document Analysis and Recognition (ICDAR'05), pp. 116–120 (2005)
12. Chen, S.: Document Layout Analysis Using Recursive Morphological Transforms. University of Washington, Seattle, WA, UMI Order No. GAX96-09607 (1996)
13. Bloomberg, D.S.: Multiresolution Morphological Approach to Document Image Analysis (1991)
14. Bukhari, S.S., Shafait, F., Breuel, T.M.: Improved document image segmentation algorithm using multiresolution morphology. In: IS&T/SPIE Electronic Imaging, p. 78740D (2011)
15. Nagy, G., Seth, S., Viswanathan, M.: A prototype document image analysis system for technical journals. Computer (Long Beach, CA). **25**(7), 10–22 (1992)

16. Ha, J., Haralick, R.M., Phillips, I.T.: Document page decomposition by the bounding-box project. In: Proceedings of 3rd International Conference on Document Analysis and Recognition, vol. 2, pp. 1119–1122 (1995)

17. Vasilopoulos, N., Kavallieratou, E.: Complex layout analysis based on contour classification and morphological operations. Eng. Appl. Artif. Intell. **65**, 220–229 (2017)

18. Normand, N., Viard-Gaudin, C.: A background based adaptive page segmentation algorithm. In: Proceedings of 3rd International Conference on Document Analysis and Recognition, vol. 1, pp. 138–141 (1995)

19. Kise, K., Yanagida, O., Takamatsu, S.: Page segmentation based on thinning of background. In: Proceedings of 13th International Conference on Pattern Recognition, vol. 3, pp. 788–792 (1996)

20. Kise, K., Sato, A., Iwata, M.: Segmentation of page images using the area Voronoi diagram. Comput. Vis. Image Underst. **70**(3), 370–382 (1998)

21. Kaur, R.P., Jindal, M.K., Kumar, M.: Text and graphics segmentation of newspapers printed in Gurmukhi script: a hybrid approach. Vis. Comput. **37**, 1–23 (2020)

22. Antonacopoulos, A., Ritchings, R.T.: Flexible page segmentation using the background. In: Proceedings of the 12th IAPR International Conference on Pattern Recognition, Vol. 3-Conference C: Signal Processing (Cat. No. 94CH3440-5), vol. 2, pp. 339–344 (1994)

23. Pavlidis, T., Zhou, J.: Page segmentation and classification. CVGIP Graph. Model. Image Process. **54**(6), 484–496 (1992)

24. Mehri, M., Nayef, N., Héroux, P., Gomez-Krämer, P., Mullot, R.: Learning texture features for enhancement and segmentation of historical document images. In: Proceedings of the 3rd International Workshop on Historical Document Imaging and Processing, pp. 47–54 (2015)

25. Chen, K., Seuret, M., Liwicki, M., Hennebert, J., Ingold, R.: Page segmentation of historical document images with convolutional autoencoders. In: 2015 13th International Conference on Document Analysis and Recognition (ICDAR), pp. 1011–1015 (2015)

26. Coquenet, D., Chatelain, C., Paquet, T.: DAN: a segmentation-free document attention network for handwritten document recognition. IEEE Trans. Pattern Anal. Mach. Intell., 1–17 (2023)

27. Yang, X., Yumer, E., Asente, P., Kraley, M., Kifer, D., Lee Giles, C.: Learning to extract semantic structure from documents using multimodal fully convolutional neural networks. In: Proceedings of the IEEE Conference on Computer Vision and Pattern Recognition, pp. 5315–5324 (2017)

28. Wick, C., Puppe, F.: Fully convolutional neural networks for page segmentation of historical document images. In: 2018 13th IAPR International Workshop on Document Analysis Systems (DAS), pp. 287–292 (2018)

29. Oliveira, S.A., Seguin, B., Kaplan, F.: dhSegment: a generic deep-learning approach for document segmentation. In: 2018 16th International Conference on Frontiers in Handwriting Recognition (ICFHR), pp. 7–12 (2018)

30. Chen, K., Seuret, M., Hennebert, J., Ingold, R.: Convolutional neural networks for page segmentation of historical document images. In: 2017 14th IAPR International Conference on Document Analysis and Recognition (ICDAR), vol. 1, pp. 965–970 (2017)

31. Zhang, H., Xu, C., Shi, C., Bi, H., Li, Y., Mian, S.: HSCA-net: a hybrid spatial-channel attention network in multiscale feature pyramid for document layout analysis. J. Artif. Intell. Technol. **3**(1), 10–17 (2023)

32. Fletcher, L.A., Kasturi, R.: A robust algorithm for text string separation from mixed text/graphics images. Pattern Anal. Mach. Intell. IEEE Trans. **10**(6), 910–918 (1988)

33. Zlatopolsky, A.A.: Automated document segmentation. Pattern Recogn. Lett. **15**(7), 699–704 (1994)

34. Simon, A., Pret, J.-C., Johnson, A.P.: A fast algorithm for bottom-up document layout analysis. IEEE Trans. Pattern Anal. Mach. Intell. **19**(3), 273–277 (1997)

35. Bukhari, S.S., Azawi, A., Ali, M.I., Shafait, F., Breuel, T.M.: Document image segmentation using discriminative learning over connected components. In: Proceedings of the 9th IAPR International Workshop on Document Analysis Systems, pp. 183–190 (2010)

36. Zagoris, K., Chatzichristofis, S.A., Papamarkos, N.: Text localization using standard deviation analysis of structure elements and support vector machines. EURASIP J. Adv. Signal Process. **2011**(1), 1–12 (2011)

37. Tran, T.-A., Na, I.-S., Kim, S.-H.: Separation of text and non-text in document layout analysis using a recursive filter. KSII Trans. Internet Inf. Syst. **9**(10), 4072–4091 (2015)

38. Tran, T.A., Na, I.-S., Kim, S.-H.: Hybrid page segmentation using multilevel homogeneity structure. In: Proceedings of the 9th International Conference on Ubiquitous Information Management and Communication, pp. 1–6 (2015)

39. Le, V.P., Nayef, N., Visani, M., Ogier, J.-M., De Tran, C.: Text and non-text segmentation based on connected component features. In: 2015 13th International Conference on Document Analysis and Recognition (ICDAR), pp. 1096–1100 (2015)

40. Oyedotun, O.K., Khashman, A.: Document segmentation using textural features summarization and feedforward neural network. Appl. Intell. **45**, 1–15 (2016)

41. Lin, M.W., Tapamo, J.-R., Ndovie, B.: A texture-based method for document segmentation and classification. S. Afr. Comput. J. **36**(1), 49–56 (2006)

42. Journet, N., Eglin, V., Ramel, J.-Y., Mullot, R.: Text/graphic labelling of ancient printed documents. In: Eighth International Conference on Document Analysis and Recognition (ICDAR'05), pp. 1010–1014 (2005)

43. Acharyya, M., Kundu, M.K.: Multiscale segmentation of document images using m-band wavelets. In: International Conference on Computer Analysis of Images and Patterns, pp. 510–517 (2001)

44. Tran, T.A., Na, I.S., Kim, S.H.: Page segmentation using minimum homogeneity algorithm and adaptive mathematical morphology. Int. J. Doc. Anal. Recognit. **19**(3), 191–209 (2016)

45. Zhong, X., Tang, J., Yepes, A.J.: Publaynet: largest dataset ever for document layout analysis. In: 2019 International Conference on Document Analysis and Recognition (ICDAR), pp. 1015–1022 (2019)

46. Soto, C., Yoo, S.: Visual detection with context for document layout analysis. In: Proceedings of the 2019 Conference on Empirical Methods in Natural Language Processing and the 9th International Joint Conference on Natural Language Processing (EMNLP-IJCNLP), pp. 3464–3470 (2019)

47. Antonacopoulos, A., Clausner, C., Papadopoulos, C., Pletschacher, S.: ICDAR2015 competition on recognition of documents with complex layouts—RDCL2015. In: 2015 13th International Conference on Document Analysis and Recognition (ICDAR), pp. 1151–1155 (2015)

48. Clausner, C., Antonacopoulos, A., Pletschacher, S.: ICDAR2017 competition on recognition of documents with complex layouts—RDCL2017. In: Proceedings of the International Conference on Document Analysis and Recognition, ICDAR, vol. 1, pp. 1404–1410 (2017). https://doi.org/10.1109/ICDAR.2017.229

49. Antonacopoulos, A., Pletschacher, S., Bridson, D., Papadopoulos, C.: ICDAR 2009 page segmentation competition. In: 2009 10th International Conference on Document Analysis and Recognition, pp. 1370–1374 (2009)

50. Lu, T., Dooms, A.: Probabilistic homogeneity for document image segmentation. Pattern Recognit. **109**, 107591 (2021)

51. Clausner, S.P.C., Antonacopoulos, A.: ICDAR2019 competition on recognition of documents with complex layouts—RDCL2019. In: Proceedings of the 15th International Conference on Document Analysis and Recognition (ICDAR2019), pp. 1521–1526 (2019)

52. Li, M., et al.: DocBank: A Benchmark Dataset for Document Layout Analysis. arXiv Prepr. arXiv2006.01038 (2020)

53. UCI Machine Learning Repository. http://archive.ics.uci.edu/ml/datasets/Newspaper+and+magazine+images+segmentation+dataset# (2013)

54. UW-III English/Technical Document Image Database. http://isis-data.science.uva.nl/events/dlia//datasets/uwash3.html

55. The MediaTeam Document Database. http://www.mediateam.oulu.fi/downloads/MTDB/download2.html

56. Papadopoulos, C., Pletschacher, S., Clausner, C., Antonacopoulos, A.: The IMPACT dataset of historical document images. In: Proceedings of the 2nd International Workshop on Historical Document Imaging and Processing, pp. 123–130 (2013)
57. Pletschacher, S., Antonacopoulos, A.: The PAGE (page analysis and ground-truth elements) format framework. In: 2010 20th International Conference on Pattern Recognition, pp. 257–260 (2010)
58. Saad, R.S.M., Elanwar, R.I., Kader, N.S.A., Mashali, S., Betke, M.: BCE-Arabic-v1 dataset: towards interpreting Arabic document images for people with visual impairments. In: Proceedings of the 9th ACM International Conference on PErvasive Technologies Related to Assistive Environments, pp. 1–8 (2016)
59. Tran, T.A., Oh, K., Na, I.-S., Lee, G.-S., Yang, H.-J., Kim, S.-H.: A robust system for document layout analysis using multilevel homogeneity structure. Expert Syst. Appl. **85**, 99–113 (2017)
60. Shafait, F., Smith, R.: Table detection in heterogeneous documents. In: Proceedings of the 9th IAPR International Workshop on Document Analysis Systems, pp. 65–72 (2010)
61. DIOTEK Mobile Software Development Company. http://www.diotek.com/eng/
62. Antonacopoulos, A., Bridson, D., Papadopoulos, C., Pletschacher, S.: A realistic dataset for performance evaluation of document layout analysis. In: 10th International Conference on Document Analysis and Recognition, 2009. ICDAR'09, pp. 296–300 (2009)

Chapter 4
Document Region Classification

Abstract After region segmentation, it is necessary to identify the functional labels of the segmented regions. This is because OCR engines can only consider text regions. Presence of non-text may degrade their performance. On the other hand, different non-texts follow different ways to present information. Therefore, these require separate processing. To satisfy this need, classification of the segmented regions in various types like text, image, table, chart, separator, etc. is essential. In this chapter, different methods developed for document region classification are discussed along with their pros and cons. This chapter also discusses the standard datasets and metrices developed for performance evaluation of these methods.

Keywords Text non-text separation · Table processing · Chart processing · Semantic segmentation · Deep learning · CNN · MRF · VGG-Net · Inception Net · HMM

The objective of document region classification (DRC) is to classify the segmented regions of a document image as per their functional role. Let, D be a segmented document with n constituent regions such as $D = \{r_i | 1 \leq i \leq n\}$, and U is a universal set of different region types, then DCR can be expressed as a mapping between D and U, i.e. $f_{DCR} : D \rightarrow U$. DCR facilitates many further processing of document images. These include content retrieval, document classification, indexing, automatic interpretation, effective searching and many more. In this chapter, many recent region classification methods are discussed along with the earlier ones but before that in the next section, different categories of document regions are presented.

4.1 Different Types of Document Regions

As said, any document image can be considered as an organized collection of various types of regions. These can broadly be categorized as text and non-text regions. A text region further can be of type paragraph, caption, title, header, footer, drop-capital, etc. [1]. Similarly, a non-text region can be an image, graph, separator,

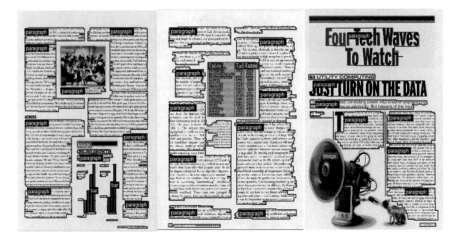

Fig. 4.1 Various regions present in a document image. Blue indicates text regions, Green represent image regions, regions with black border are charts, regions with brown border are tablets, and pink border are separators

margin, seal, etc. Another very commonly found region is table [2]. Figure 4.1 depicts various regions of a document image.

4.2   Different Document Region Classification Methods

In literature, a wide range of method is proposed to address the document region classification problem. Many of these methods mostly identify the regions as text or non-text [3]. For the classification or detection of various types of non-text or text regions, researches often develop type specific methods [4, 5]. Therefore, in this section, first the methods proposed for text non-text separation are discussed and then methods developed for specific types of non-text regions are described.

4.2.1   Methods for Text/Non-text Classification

Identification of text regions separately from the non-text region is essential to facilitate text processing in order to support many document processing applications. However, the complexity of this task varies based on the type of document under consideration. For example, in online documents which are generated using the device like, ipad, electronic white board, the writing styles of various writers result in huge shape variation among the components. This causes ambiguity between these two categories of regions [6]. Similar, difficulties can be observed in offline hand-written documents. Attachment of text with non-text occurs frequently in these

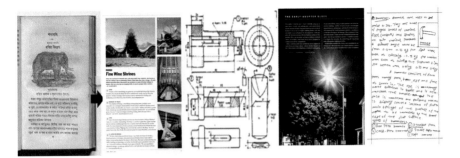

Fig. 4.2 Shows some sample images of different categories of documents

documents which is a major concern in text non-text separation. Localization of text regions in scene images has many challenges caused by *divers environmental conditions* (i.e. uneven light, complex scene), *bad image acquisition* (i.e. blurring, perspective distortion), *complex text* [7], etc. Similarly, in web images, low resolution of the web content, presence of too many graphical objects pose the difficulty in the way of classification. Beside these document types, text non-text separation in documents like newspaper, magazine, magazine/book covers, engineering drawings, maps, historical documents is also a complex task. This is because of the wide verity present in the shape, size, and orientation of the components, background color variations, overlapping of text and non-text regions, document noise, etc. [3]. Figure 4.2, presents documents from different categories to ease the understanding of challenges present in text non-text separation.

However, in the present scope of the discussion, different methods developed for text non-text separation are discussed irrespective of the document types for which these are developed. Here, the methods are categorized as *non-machine learning based methods*, *shallow learning based methods*, and *deep learning based methods*.

Non-machine Learning Based Methods

Most of the earlier methods developed for text non-text separation belong to this category. These methods use a pre-defined set of rules for the classification task. For example, authors in [8], represent every region of a document image using "white tiles" and classify these regions based on a set of rules. A white tile is a largest rectangle that can be used to represent widest white background area present in between the foreground elements. Authors assume, in text regions, most of the white tiles will be narrow whereas, in non-text regions, wide white tiles will be present in majority. Although this method is skew independent but may suffer for documents with overlapping regions and color background. A similar method is reported in [9], where authors use run length smearing based features for block classification.

There are some methods of this category which estimate the features from connected components. On such method is reported in [10], here authors estimate the component frequency on the basis of height, area covered by the components and white pixels. After that they use a set of rules to distinguish text regions from non-text regions. Authors in [11], initially, filter the extracted components based on various geometrical information (like size, aspect ratio, shape), positional information of the component, and the number of small components contained by the bounding box of the component under consideration. Authors name it as heuristic filtering. In the next level, a cursive filter is used to detect the homogeneous component, background is analysed, and statistical features are extracted to separated remaining text and non-text components. Recently, in [12, 13], authors follow a similar approach to classify the connected components present in Bangla filled in form images as text or non-text. A wide range of methods developed to separate text part from the non-text part in engineering drawings follow similar approach at the initial stage [14].

Few very recent DLA methods used similar approach for region classification and obtained very impressive result. DSPH [15] is one of these methods which won the RDCL 2019 competition on recognition of documents with complex layouts. This method uses a probabilistic framework to measure homogeneity of a region for region classification. MHS [16–18] which secures the second position in RDCL 2019 and won RDCL 2017 competition, also use the concept of minimum and multilevel homogeneity for the said task. The method name BINYAS [19] and ABCD [20] which appear in third position in RDCL 2019 competition and first position in REID 2019 competition on recognition of early Indian printed documents respectively, also extract the geometrical properties of the regions and based on some developed rules classify these as text or non-text.

Despite of their excellent performance it can be argued that it is always difficult to device a generalized set of rules or generalize thresholds for a wide range of documents. Besides, most of these methods consider binarized image as input. Therefore, bad binarization of noisy document may severely affect their performance.

Shallow Learning Based Methods

Methods that come under the purview of this category first, extract different statistical [21], and shape based features from the segmented regions and then use different machine learning algorithms to classify these [22].

However, among different machine learning algorithms Artificial neural networks (ANN) are most popularly used for this purpose. For example, authors in [23], use Self organize feature map (SOM) to extract local features from the segmented regions to classify these as text or non-text. In [24], histogram of oriented gradient (HOG) features are used to represent the regions in the features space and finally classified using Multi-layer perceptron (MLP). Besides these, a good number of methods use gradient shape based features [25], Scale invariant feature transform based features [26, 27] to classify the segmented blocks using ANN. Authors in [28], explore the performance of four different ANNs which include, SOM, Probabilistic neural network, MLP, and RBF network in identifying regions as text or non-text. Many researchers prefer to use different types of classifiers to obtain the final classification result like authors in [29] use Naïve Bayes (NB), MLP, AdaBoost, Random forest (RF), and Support Vector Machine (SVM) to obtain the final classification result. In this work, a script invariant feature descriptor is designed based on the area occupancy profile of the equidistant pixels to represent the regions in the feature space. Recently, authors in [30], extract texture based features like Local binary pattern (LBP) and local ternary pattern (LTP) features to represent the regions extracted from the document pages of different RDCL datasets. Authors use a modified version of binary particle swarm optimization (BPSO) based feature selection algorithm to eliminate the redundant features. To classify the regions based on the truncated feature set, they use RF classifier. This method repots impressive result in the said dataset. Although these methods obtain good results, their performance heavily depends on the region segmentation result. Any erroneous segmentation that results in regions with both text and non-text may severely affect the performance of these methods.

To address this issue, a significant number of methods extract features from connected components (CCs) and classify these using different classifiers as text or non-text. CC based methods also extract different statistical and shape based features from CCs. One such method is reported in [31], where authors extract shape and context information from the CCs and use MLP to classify these based on the extracted features. Here authors assume the context of text components will always be structured. However, this will not be valid for text CCs that appear around non-texts. This may lead to misclassification of such CCs. In literature, many methods that are reported to perform text non-text separation in handwritten documents, compute different variants of LBP features from extracted CCs. One such method is reported in [32]. In this work, authors extract CCs from handwritten lab copies and then compute different LBP based features from the extracted CCs. Additionally, for feature extraction purpose along with five different variants of LBP, authors introduce a new variant, named robust and uniform LBP (RULBP). Finally, the classification of the components based on these feature descriptors is done using five different classifiers which include Naïve Bayes (NB), MLP, K-nearest neighbor (K-NN), RF, and Support Vector Machine (SVM). Among these, the combination of RULBP feature descriptor and RF classifier obtains the best result. Similar approach is followed in [33, 34]. Authors in [34], use coalition game based feature selection algorithm to identify the most informative regions of an input image so that the features extracted from the non-informative regions can be eliminated. This reduces the training time and increases the classification accuracy by eliminating the redundant features.

Deep Learning Based Methods

In recent days, Deep learning (DL) models are becoming the state-of-the-art in the domain of DLA. Methods that come under this category mostly perform pixel level classifications. For pixel level classification FCNNs are commonly preferred because these are comparably light weight end-to-end method [35]. One such method is reported in [36], where authors develop a unified multimodal FCNN consisting of encoder, decoder and bridge to initially identify the text and non-text part by classifying pixels and then perform semantic segmentation. Authors in [37] develop a CNN based model followed by task specific post processing blocks to address the said problem in historical documents. A similar work reported in [38] for historical documents. Here a FCNN with encoder and decoder is used for pixel classification. Authors here use binarized version of the input image to refine the output of the said model. In contrast to these complex models, a relatively simple CNN model with one convolution layer is proposed in [39] for pixel classification. The model obtains comparable result. Recently, authors in [40] develop a Generative adversarial network to classify the text and non-text pixels present in a touching component with both text and non-text part. Authors validate the effectiveness of their work by performing OCR on the components before segmentation and after segmentation. It is observed, this segmentation step significantly enhances the recognition accuracy of the OCR engine. However, in few cases, this model generates some artifacts at the background of the segmented images.

There are some methods that use deep models to classify document blocks or regions directly as text or non-text. In [41], a simple CNN model inspired by LeNet is developed to address the said problem. This model takes cropped region images as input to identify it as text or non-text. Object detection models like Fast-Region based CNN (F-RCNN) and Mask-RCNN (M-RCNN) are also used in many recent methods developed for automatic region detection and classification. F-RCNN model has two steps; the first one generates candidate regions named as Region proposal network (RPN) and the second one extracts features to classify these candidate regions. The second stage outputs bounding box offset of the candidate regions and their corresponding classes. On the other hand, M-RCNN is an extended version of F-RCNN. This model additionally outputs a binary mask for each candidate regions. One such method is reported in [42], where authors apply both F-RCNN and M-RCNN to classify document regions. These models obtain impressive result on standard datasets like ICDAR 2013. In [43], authors only use F-RCNN for the said purpose and obtain good result. Due to region level classification, these methods may also suffer for documents where the text non-text components follow a dense arrangement. Because here any detected candidate region may consist of both text and non-text. As these methods are classifying a whole region as text or non-text, this may lead to erroneous output. There are some works recently reported in the literature which follow hybrid approach. One such is [44], where authors use four base line models which include; BERT a transformer based pre-trained model which was trained on a large text corpus, RoBERTa an extended version of BERT, LayoutLM is also a pre-trained language model which models both text and the layout, and finally F-RCNN. In this work, the first three models are applied at the token level and F-RCNN is used at the region level.

4.2.2   Methods for Text Region Classification

Type of text regions in a document image may vary from one document type to another. For newspapers these can be paragraph, title, header, footer, drop-capital, etc (see Fig. 4.3). In scientific articles, author information, equation, sub-heading, etc. are additionally found. For envelops or letters, these can sender and receiver address, paragraph, date, place, etc. Meany researchers develop separate module for classifying the identified text regions in different logical classes. One such method is reported in [1] where, authors extracts CCs from the text regions to classify it as title, header, footer, and drop-capital. To do that, they extract shape based, positional and contextual information from CCs for their logical labelling. Authors define a set of rules for the final classification. A similar work can be found in [16, 18]. In recent methods, these text classes are considered along with the non-text classes during the classifications of CCs or regions. Therefore, no separate module is used for logical labelling of text. However, this may increase the complexity of the labelling task but the use of deep models makes it possible. For example, in [45], authors develop a hybrid spatial-channel attention model (HSCA-Net) with spatial and channel attention module and lateral attention connection for region segmentation and classification. Here the channel attention module dynamically alters the channel feature responses based on the contribution of the features of each channel whereas the spatial attention module guides the CNNs to emphasise on the informative contents and capture the inter-region relationship among the page regions to extract the global context. The lateral attention connection is used to include spatial and channel attention module in multiscale feature pyramid network and to retain the original features. This model classifies regions as text, title list, table, formula, caption, table

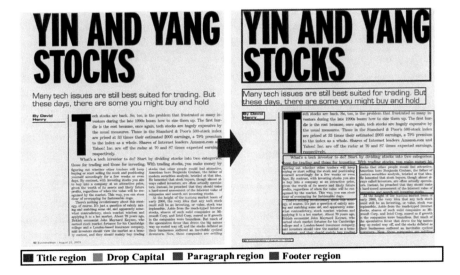

Fig. 4.3   Illustrates different classes of text regions

caption, reference, abstract, authors, figure, map, etc. Similarly, in [42] authors identify text regions like text, title, lists along with table and figure using two object detection based models F-RCNN and M-RCNN. In [43] also F-RCNN model is used to identify title, author, abstract, body, table caption and figure caption in addition to table and figure. Authors in [44] use four baseline models BERT, RoBERTa, LayoutLM, and F-RCNN for detecting abstract, author, caption, equation, figure, footer, list, paragraph, reference, section, table, title etc.

4.2.3   Methods for Non-text Classification

Visually rich document images commonly posses' non-texts like charts, tables, maps, images, separators, etc. However, in historical documents, different drawings, seals, and page decorative are commonly found. Most of the earlier methods and few recent methods [16, 18, 19] use shape based information like aspect ratio, and pixel density based information to identify non-texts like separators, bullets, images, etc. As mentioned in the previous section resent methods [42–44] do not have separate module for non-text classification rather try to identify different text and non-text classes in one go. However, a significant number of methods are found in the literature for table and chart detection. In this section, the recent advancements done in these domains are presented in a lucid way.

Table Detection

Tables are the popular means of representing structured data. Therefore, widely used in financial documents, scientific documents, purchase order and many more. Processing of tables involves *table detection* and *structure recognition*. Table detection is to identify the table boundary in the input image, whereas, structure recognition is to extract the row, column, and the cell structure of the detected table (see Fig. 4.4). However, there are multiple challenges to be taken care of in order to perform these tasks. One of these is the diversity present in the structure of a table. A table can be fully closed, semi-closed, and even in some documents tables with no border can be found. Besides, a cell of a table may contain a single character, a paragraph, a formula, a figure, etc. Often other document objects like figure, flowchart, etc. share similar textural properties and thus makes the table detection complex. Tables having other tabular objects embedded within it are also difficult to process.

Most of the earlier methods developed for table detections are rule based [46]. These methods usually rely on visual evidences [19], keywords, formal templates, etc. These methods require huge manual effort to develop the heuristic and set the parameters. However to deal with this issue many shallow learning based methods are introduced [47]. These methods identify the table locations using different classifiers based on different hand-crafted features. Although these methods

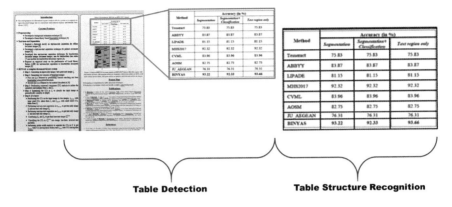

Table Detection   **Table Structure Recognition**

Fig. 4.4 Illustrates table detection and structure recognition

bring significant improvement in table detection accuracy, the use of hand-crafted features limit their ability of generalization. Later, many deep learning based methods using CNN based model are developed for the said task, which produce state-of-the-art results [4]. One group of these methods follow top-down object detection framework like RCNN to address the problem of table detection [48, 49]. RCNN generates candidate regions using a process named selective search and then the candidate regions are sent to AlexNet for deep feature extraction. Finally, it uses SVM classifier to perform the classification. RCNN performs a forward CNN pass for each candidate region and it generates around 2000 regions per image. Therefore, slow in nature. However, later advanced models like F-RCNN, M-RCNN, Faster RCNN, YOLO (You Only Look Once) object detection models are used for table detection [42, 43, 50, 51]. F-RCNN improves the processing time by sharing the convolution operation among the candidate regions. On the other hand, Faster-RCNN uses CNN model to generate candidate regions and thus makes the region generation a part of CNN training and prediction. YOLO eliminates the candidate region generation part. Instead, it uses a CNN to identify the region bounding boxes and predict the classes in one evaluation directly from the input image. Some researchers show that the performance of these models for table detection can further be improved by adding some pre-processing like colorization and dilation, distance transforms, etc. to enhance the input image [52]. Authors in [53, 54] use deformable convolution in their object detection models to improve the accuracy. These models obtain state-of-the-art performance in many standard datasets. However, the computational complexity and space complexity of these models are high. Another category of methods considers the problem of table detection as semantic segmentation problem and use the existing models for this purpose. These methods first generate pixel level mask and then combine the pixels to form the region. One such method is reported in [36], where a multimodal FCN is used for the table and other document object detection. Here authors consider both the visual and linguistic features to obtain improved accuracy. In [55], multi-scale multi-tasking FCN is developed for pixel level mask prediction which is then refined using CRF model and post processed to detect table region. There are some methods [44, 56] available in the literature which consider table detection as a graph labelling problem. These methods consider a document as a graph where different document objects are considered as node and edges are the relationships between two neighbouring nodes. Authors in [56] generate a neighbourhood graph where the text objects are considered as nodes. Then a Graph neural network (GNN) is used to classify the nodes and the edges. After that the connected subgraphs formed with the nodes classified as table are extracted as table. Authors in [44] perform the work at the token level and use three text processing models and one object detection model to address the problem. Although, these methods obtain good results, but these require accurate text or token bounding box information as additional input.

Table structure recognition involves in extracting the exact structure of the detected table, i.e. bounding boxes of rows, columns, span of rows and columns or cellular structure, etc. Just like table detection, a significant number of methods using deep models are developed for table structure recognition. One group of these methods use either semantic segmentation or object detection or both to extract rows and columns. Authors in [57], use two FCN based semantic segmentation models to detect rows and columns. The output is farther post-processed to handle issues like false detection, severed and connected structures. However, this method suffers badly for the table structures with large blank spaces [58, 59]. Authors in [58, 59], modify the pooling operation of their FCN models to pool features along rows and columns of pixels in some intermediate feature maps so that the row and column recognition accuracy can be improved even for the tables with large blank space. Authors in [60] use Gate recurrent unit network (GRUN) to extract the rows and columns from a input table image by scanning this from left-to-right, and top-to-bottom. Some methods from this category use object detection models to locate rows and columns. One such method can be found in [61], here, authors use three object detection models namely, deformable F-RFCN, deformable Fast RCNN, and deformable FPN to identify the bounding boxes of the rows and columns. Similarly authors in [62] use Mask-RCNN for the said purpose. Many of these object detection methods only focus on the basic grid structure of the table, did not address the issue of merged cell [4]. However, there are few methods like one reported in [59], in addition to structure detection, deals with merged cell issue. Despite of their impressive performance, these methods cannot be directly adopted for the distorted or curved table images. This is because of their implicit assumption that tables are axis-aligned. Another group of methods for table structure recognition rely on image-to-markup generation models to convert the detected table to the targeted markup representation which can describe its complete cellular structure. To do that mostly encoder-decoder architectures are used along with large datasets [63, 64]. These methods require large training data and suffer when the table structure is complex and big. There are also some graph based methods [65, 66] which consider the detected table as a graph with cell contents as nodes. These methods then use GNN to predict whether two nodes are from same cell, or row, or column. Major issue with these methods is these assume the bounding box information of the cell content is available. However, extraction of exact coordinates of the cell contents from table images is often very difficult. Later, some methods [67, 68] are developed to deal with this issue. These methods first identify the bounding box of the cell content and then group them to form rows and columns. These methods also fail to deal with tables having many empty cells and distorted structure.

Chart Processing

Charts are very popular and effective way of presenting information. In contemporary documents presence of charts are common. Processing of these charts is essential to extract the information these carry. However, processing of charts involves many steps like *Chart extraction, multi-panel chart processing, chart classification, chart data extraction* and *chart description generation, etc.* [5, 69].

To extract charts from an input document image, initially the location of the charts and the respective captions are identified and then linked [69]. Most of the earlier methods involve in chart and caption region detection consider heuristics for region segmentation and classification [70]. Later many deep model based methods are developed to address the same [71, 72]. Although, these models obtain better result, it is observed deep models with saliency based attention module suffer for charts having large empty space [72]. In literature, many methods are developed that consider vector based documents as input [73, 74]. These methods analyse the operators to extract the figures. As other operators often get mixed with figure related vector graphics without any demarcation, it is very challenging to extract figures from such documents [75]. Methods developed for linking figures with appropriate captions take both figure candidates and caption candidates to generate the linked pairs of caption and figure. These methods use various heuristics related to layout and others like, captions in documents appear close to the corresponding figure or in the same pages, the geometric information of the caption, etc. to assign cost to pairing a caption with a figure. In next step all such candidate pairs are analysed based on cost to identify the best pair using greedy approach [76] or minimize the cost by identifying the optimal pair using Hungarian method [77]. For multi-panel figures, often single caption is associated with multiple figure candidates [78].

In contemporary document images, figures with multiple charts as sub-figures are common. These are called multi-panel figures [69]. Major issues with multi-panel figures is the shared regions among the panels which needs a special care. Any common segmentation algorithm may result in over segmentation [79]. Segmentation of multi-panel figures or charts has several steps. The very first step is to identify the multi-panel figures separately from the single panel figures through classification. This is to select the appropriate segmentation algorithm and minimize the risk of over segmentation [80]. The next most important step before segmentation is the analysis of the paired caption to get the number of sub-figures or panels [69]. This is to validate the segmentation result [81]. This is generally done by identifying the delimiters from the caption string based on some heuristics [81], so that the sub-captions can also be separated and can be linked to the respective sub-panels. The labels for the sub-panels are generally embedded in the figure itself. The detection and recognition of such labels are also necessary for correct mapping of sub-captions and panels. Most of the recent methods for label detection are deep model based [82] or use path based classifier [83]. These detected labels are then recognised and classified in some common panel labels [83]. For this purpose, many methods are introduced from heuristic based [76] to MRF [84], CNN [82] based. However, these methods often suffer due to low image quality and uneven layout [84]. Once the labels are detected and recognized, the next step is to segment the panel. For panel segmentation also, a large number of methods are reported in the literature [69]. Just like other domains, most of the recent panel segmentation methods follow object detection approach [85, 86]. However, these methods may suffer from region overlapping issue in absence of constrain. Some specialized CNN models also available in the literature which try to model figure layout, perform comparably better [86]. After segmentation, parsing of the subpanels are performed to produce the final list of subpanels and related sub captions. In this step, any mismatch between the estimated number of sub-figures before segmentation and the number of generated sub-figures after segmentation is addressed either by clustering panel candidates [87], or by heuristics [81].

The next most important step in chart processing is chart type classification. Most of the earlier methods for the said task is shallow learning based or model based [88–90]. These methods initially extract various features from the chart images and based on that classify these in one of many chart types. To do that, for each chart type separate models are prepared. In a recent survey [5], the features extracted from chart images are clustered in three groups; low-level, middle-level, and higher-level. In some research work, it is shown that the low-level features perform well in chart classification whenever used with Hough transform [91], and HMM [92]. Middle-level features which tries to capture the shape information mostly using HOG (Histogram of oriented gradient), or SIFT (Scale invariant feature transform) descriptors [93, 94] are generally used with multiple instance learning [95]. High-level features are mostly applied with SVM (Support vector machine) and obtain state-of-the-art results, even for small datasets [96]. Recent methods for char classification rely on Neural network based models like CNN and perform significantly better than the earlier methods. For example, the performances of Inception-V3, ResNet, and VGG-19 for chart classification task are compared with traditional model based methods which include HOG & KNN, HOG & SVM, HOG & RF, HOG & NB in a work [97]. The obtained result shows that the CNN models improve the classification accuracy more than 20% compares to the traditional models. A large number of recent methods use pretrained CNN models which are already trained with some large datasets like ImageNet for chart classification [98–100]. Here, only the final layer needs to be changed based on the number of classes considered and retrained. Commonly used CNN models for chart classifications are VGG [100, 101], Inception Net [102, 103], AlexNet [100, 104], ResNet [103, 105], GoogleNet [104, 106], MobileNet [99, 102]. The performance comparison of different CNN models in chart classification is reported in [103, 104, 107]. It is observed that the performance is these CNN models are quite similar with 5% divergence in different datasets.

After the detection of chart type, the tabular data based on which the input chart is prepared is extracted. To do that, different types of methods are developed which can broadly be grouped as semi-automatic [106, 108], and fully-automatic [104, 109]. Majority of these are fully-automatic methods which take only a chart image as input from the user, whereas, the semi-automatic methods are interactive and need manual interventions to produce results. Chart data extraction methods are limited to some specific charts like bar, pie, and line, etc. Besides most of the traditional methods are chart specific [110, 111] and are rule based. However, present methods rely on deep models. For example, in [112] authors develop a single deep model along with an RNN for the said purpose for bar chart and pie chart. Similar work can be found for different types of bar charts in [113]. Encoder-decoder architecture is also used for this task like, in [114], as encoder CNN is used whereas as decoder RNN. Although, CNNs improve chart data extraction process significantly, it is observed, when these are coupled with traditional algorithms, their performance gets improved further [5].

The final stage in this pipeline is chart description generation. The objective of this step is to generate a text description of the input chart. Just like the previous stage most of the developed methods for this purpose are limited to some specific charts. One group of these methods [102, 110] aims to generate summery for the input chart, whereas the other group aims to identify the message, the input chart intended to deliver [89, 115, 116]. The first one generates the information contained by the input chart, whereas the second one tries to capture the human perception and understanding after first time observing the chart. A more detailed discussion on chart description generation can be found in the survey paper by Bajić and Job [5].

4.3 Evaluation Techniques

As said, for evaluation of methods both labelled dataset and performance or evaluation metrics are required. In this section first, different standard datasets available for evaluation will be discussed and then different evaluation metrices will be presented.

4.3.1 Standard Datasets

For the evaluation of methods developed for text non-text separation many standard datasets are made available. For example, the Newspaper and Magazine image segmentation database [117] available under UCI Machine Learning Repository, University of Washington (UW) Document Image Database. This dataset contains 101 page images in Russian language and free for public access. The MediaTeam

Oulu Document Database [118] which possess 512 English document images. ICDAR Page segmentation Datasets [119–121], etc. Most of these databases consist of annotated images of pages from the newspapers and the magazines. Images containing chemical formulas, engineering drawings and so on are included in UW-III English/Technical Document Image Database [122]. Images containing maps, engineering drawings, business cards, forms are available with MediaTeam Oulu Document Database. Apart from these printed document databases, few *online* handwritten document databases, such as IAMonDo database [123], Kondate [124], are also available for testing text non-text separation methodologies. Due to availability of stroke information in these databases, various statistical models such as Markov Random Field (MRF) [125], Conditional Random Field (CRF) [126] are employed for modeling of contextual relationships between strokes in separating text from non-text strokes. Some archives of old documents, like, "Bichitra" [127], "IHP Collection" [128], contain handwritten non-text segments together with text. But these archives have no annotations. Thus, text non-text separation algorithms cannot be trained and tested directly using these datasets.

The said datasets can also be used for text classification as well as non-text classifications. However, in this subsection, some standard datasets for table processing and chart processing are also discussed separately. There are many standard datasets available for table detection and table structure recognition. For example, PubLayNet [42] dataset which contains 335,703 training images, 11,245 images for validation and 11,405 test images. Although, this dataset is developed for DLA, it can also be used to evaluate methods developed for table detection. Another popular dataset for table detection is cTDaR TrackA [129] dataset. One subset of this dataset possesses 600 training images and 199 test images of hand drawn tables and text taken from historical documents, whereas another subset includes 600 training and 240 test images of tables from contemporary documents. The IIIT-AR-13K [130] dataset with 9333 training, 1955 validation, and 2120 test images was developed for detection of graphical objects in annual reports but can be used for table detection task. As said, there are many standard datasets made available for table structure detection. One such is SciSTR [68] dataset with 12,000 images for training and 3000 images for testing. All these images are taken from scientific papers. A dataset with large number of three-lines table images named PubTabNet [131] is also available for the said task. This dataset has total 500,777 images for training, 9115 images for validation and 9138 testing images.

For every step of chart mining a good number of datasets are made available to evaluate the performance [69]. For example, for training and validation of figure extraction methods, there exist DeepFigures [71] dataset which contains around 5,500,000 articles, the CS-Large [77] dataset which possesses 346 articles. Many such datasets are also available for the said task [76]. For the single and multi panel figure classification ImageCLEF datasets [132, 133] having more than 40,000 images are available. For figure type classification DocFigure [134] dataset with 33,070 images can be used. For chart type classification, VIEW [96], ReVision [98], DeepChart [135], CHART 2019—PMC [136] and many others can be found [104]. For chart data extraction task, SIGHT datasets [137, 138], CHIME-R [139],

DVQA [140], CHART 2019 [136], etc. are exist for training and validation. Finally, for chart text detection, recognition and roll classification also a significant number of datasets are introduced which include, ICDAR 2019 datasets [136] with around 202,950 samples images, the news dataset [98] with 475 images, the Bio-med dataset [83] with 10,642 samples and many others [98, 139].

4.3.2   Evaluation Metrices

Evaluation of text non-text separation is generally done either in qualitative way [141] or in quantitative way [30]. Qualitative assessment refers to visual examination of the outputs. Most of the earlier methods follow this approach [141]. However, the recent methods mostly rely on quantitative assessment. Most popularly used metrices are *Precision, Recall, F-measure*, and *accuracy* [3]. For table detection task, the performances are assessed using *Intersection-over-Union* (IoU) based metrices. The IoU is computed by comparing the generated object bounding box with the one in the ground truth. Generally, for performance measurement average over multiple IoU values are taken at different thresholds [4]. For table structure recognition, adjacency relation based metric used in ICDAR 2013 table competition [142] is commonly used. In this metric, a list of adjacency relations between the contents of each cell in the table and its horizontal and vertical neighbouring cells are generated. Each adjacency relation is represented using a tuple which contains the content of two cells, their direction, and the number of blank cells between these. After generation, this list is then compared with ground truth. There is another very popular metric available in the literature with the name *Tree edit distance similarity* (TEDS) based metric for table recognition [131]. TEDS metric includes both table recognition and OCR errors. However, later the TEDS metric is modified to ignore the OCR errors [143] and got the name TEDS-Struct.

There are different matrices to evaluate every step of a complete chart processing system. For example, the region based metrices like, IoU, *intersection over target box, intersection over candidate box* are used to assess the performance of figure or chart extraction methods [77]. This is done by comparing the generated result with the ground truth. For the evolution of methods developed for multi-panel chart segmentation, chart classification, generally, *Precision, Recall,* and *F-measure* metrices are used [144]. Text detection algorithms to detect text on charts are also evaluated using precision and recall metrices [145]. Recognized text from the charts are evaluated using metrics like *character error rate, word error rate, Levenshtein distance* and *Gestalt Pattern Matching* [69]. The chart data extraction methods are generally evaluated by reconstructing the chart from the extracted data [88]. This is mostly done in absence of ground truth. Recently, quantitative metrices like, precision, recall become popular due to the availability of sufficient annotated dataset. For the evaluation of chart description generation methods, word based untrained metrices like, *Bilingual evaluation understudy* (BLEU), and *Recall-oriented understudy for gisting evaluation* (ROUGE), and trained metrics like, *Bilingual evaluation understudy with representations from transformers* (BLEURT) are used [146]. BLEU and ROUGH both compare the input sentence with the reference sentence and assign score on word overlapping between these two sentences. However, BLEU is precision based metric whereas ROUGH is recall based. BLEURT is a neural network based metric. It estimates the semantic similarity between two sentences [147].

References

1. Bhowmik, S., Sarkar, R.: Classification of text regions in a document image by analyzing the properties of connected components. In: 2020 IEEE Applied Signal Processing Conference (ASPCON), pp. 36–40 (2020)
2. Bhowmik, S., Sarkar, R.: An integrated document layout analysis system. In: ICDAR2019 Doctoral Consortium, pp. 12–14 (2019). Available: http://icdar2019.org/wp-content/uploads/2019/09/LeafletDC-ICDAR19.pdf
3. Bhowmik, S., Sarkar, R., Nasipuri, M., Doermann, D.: Text and non-text separation in offline document images: a survey. Int. J. Doc. Anal. Recognit. 21(1–2), 1–20 (2018)
4. Ma, C., Lin, W., Sun, L., Huo, Q.: Robust table detection and structure recognition from heterogeneous document images. Pattern Recogn. **133**, 109006 (2023)
5. Bajić, F., Job, J.: Review of chart image detection and classification. Int. J. Doc. Anal. Recognit., 1–22 (2023). https://doi.org/10.1007/s10032-022-00424-5
6. Van Phan, T., Nakagawa, M.: Combination of global and local contexts for text/non-text classification in heterogeneous online handwritten documents. Pattern Recogn. **51**, 112–124 (2016)
7. Ye, Q., Doermann, D.: Text detection and recognition in imagery: a survey. Pattern Anal. Mach. Intell. IEEE Trans. **37**(7), 1480–1500 (2015)
8. Antonacopoulos, A., Ritchings, R.T.: Representation and classification of complex-shaped printed regions using white tiles. In: Proceedings of the 3rd International Conference on Document Analysis and Recognition, 1995, vol. 2, pp. 1132–1135 (1995)
9. Shih, F.Y., Chen, S.-S.: Adaptive document block segmentation and classification. IEEE Trans. Syst. Man, Cybern. Part B. **26**(5), 797–802 (1996)
10. Drivas, D., Amin, A.: Page segmentation and classification utilising a bottom-up approach. In: Proceedings of the 3rd International Conference on Document Analysis and Recognition, 1995, vol. 2, pp. 610–614 (1995)
11. Tran, T.-A., Na, I.-S., Kim, S.-H.: Separation of text and non-text in document layout analysis using a recursive filter. KSII Trans. Internet Inf. Syst. **9**(10), 4072–4091 (2015)
12. Bhattacharya, R., Malakar, S., Ghosh, S., Bhowmik, S., Sarkar, R.: Understanding contents of filled-in Bangla form images. Multimed. Tools Appl. **80**(3), 3529–3570 (2021). https://doi.org/10.1007/s11042-020-09751-3
13. Ghosh, S., Bhattacharya, R., Majhi, S., Bhowmik, S., Malakar, S., Sarkar, R.: Textual Content Retrieval from Filled-in Form Images. Commun. Comput. Inf. Sci. **1020**, 27–37 (2019). https://doi.org/10.1007/978-981-13-9361-7_3
14. Tombre, K., Tabbone, S., Pélissier, L., Lamiroy, B., Dosch, P.: Text/graphics separation revisited. In: Document Analysis Systems V, pp. 200–211. Springer (2002)
15. Lu, T., Dooms, A.: Probabilistic homogeneity for document image segmentation. Pattern Recognit. **109**, 107591 (2021)
16. Tran, T.A., Na, I.S., Kim, S.H.: Page segmentation using minimum homogeneity algorithm and adaptive mathematical morphology. Int. J. Doc. Anal. Recognit. **19**(3), 191–209 (2016)

17. Tran, T.A., Na, I.-S., Kim, S.-H.: Hybrid page segmentation using multilevel homogeneity structure. In: Proceedings of the 9th International Conference on Ubiquitous Information Management and Communication, pp. 1–6 (2015)

18. Tran, T.A., Oh, K., Na, I.-S., Lee, G.-S., Yang, H.-J., Kim, S.-H.: A robust system for document layout analysis using multilevel homogeneity structure. Expert Syst. Appl. **85**, 99–113 (2017)

19. Bhowmik, S., Kundu, S., Sarkar, R.: BINYAS: a complex document layout analysis system. Multimed. Tools Appl., 8471–8504 (2020). https://doi.org/10.1007/s11042-020-09832-3

20. Clausner, C., Antonacopoulos, A., Derrick, T., Pletschacher, S.: ICDAR2019 competition on recognition of early indian printed documents—REID2019. In: 2019 International Conference on Document Analysis and Recognition (ICDAR), pp. 1527–1532 (2019)

21. Oyedotun, O.K., Khashman, A.: Document segmentation using textural features summarization and feedforward neural network. Appl. Intell. **45**, 1–15 (2016)

22. Marinai, S., Gori, M., Soda, G.: Artificial neural networks for document analysis and recognition. IEEE Trans. Pattern Anal. Mach. Intell. **27**(1), 23–35 (2005)

23. Strouthopoulos, C., Papamarkos, N.: Text identification for document image analysis using a neural network. Image Vis. Comput. **16**(12–13), 879–896 (1998)

24. Sah, A.K., Bhowmik, S., Malakar, S., Sarkar, R., Kavallieratou, E., Vasilopoulos, N.: Text and non-text recognition using modified HOG descriptor. In: 2017 IEEE Calcutta Conference, CALCON 2017—Proceedings, Jan 2018, pp. 64–68 (2018). https://doi.org/10.1109/CALCON.2017.8280697

25. Diem, M., Kleber, F., Sablatnig, R.: Text classification and document layout analysis of paper fragments. In: 2011 International Conference on Document Analysis and Recognition, pp. 854–858 (2011)

26. Garz, A., Diem, M., Sablatnig, R.: Detecting text areas and decorative elements in ancient manuscripts. In: 2010 12th International Conference on Frontiers in Handwriting Recognition, pp. 176–181 (2010)

27. Wei, H., Chen, K., Ingold, R., Liwicki, M.: Hybrid feature selection for historical document layout analysis. In: 2014 14th International Conference on Frontiers in Handwriting Recognition, pp. 87–92 (2014)

28. Le, D.X., Thoma, G.R., Wechsler, H.: Classification of binary document images into textual or nontextual data blocks using neural network models. Mach. Vis. Appl. **8**, 289–304 (1995)

29. Khan, T., Mollah, A.F.: Text non-text classification based on area occupancy of equidistant pixels. Proc. Comput. Sci. **167**, 1889–1900 (2020). https://doi.org/10.1016/j.procs.2020.03.208

30. Ghosh, S., Hassan, S.K., Khan, A.H., Manna, A., Bhowmik, S., Sarkar, R.: Application of texture-based features for text non-text classification in printed document images with novel feature selection algorithm. Soft Comput. **26**(2), 891–909 (2022)

31. Bukhari, S.S., Azawi, A., Ali, M.I., Shafait, F., Breuel, T.M.: Document image segmentation using discriminative learning over connected components. In: Proceedings of the 9th IAPR International Workshop on Document Analysis Systems, pp. 183–190 (2010)

32. Ghosh, S., Lahiri, D., Bhowmik, S., Kavallieratou, E., Sarkar, R.: Text/non-text separation from handwritten document images using LBP based features: an empirical study. J. Imaging. **4**(4), 57 (2018)

33. Bhowmik, S., Sarkar, R., Nasipuri, M.: Text and non-text separation in handwritten document images using local binary pattern operator. In: Proceedings of the 1st International Conference on Intelligent Computing and Communication, pp. 507–515 (2017)

34. Ghosh, M., Ghosh, K.K., Bhowmik, S., Sarkar, R.: Coalition game based feature selection for text non-text separation in handwritten documents using LBP based features. Multimed. Tools Appl. **80**, 1–21 (2020)

35. Coquenet, D., Chatelain, C., Paquet, T.: DAN: a segmentation-free document attention network for handwritten document recognition. IEEE Trans. Pattern Anal. Mach. Intell. **99**, 1–17 (2023)

36. Yang, X., Yumer, E., Asente, P., Kraley, M., Kifer, D., Lee Giles, C.: Learning to extract semantic structure from documents using multimodal fully convolutional neural networks. In: Proceedings of the IEEE Conference on Computer Vision and Pattern Recognition, pp. 5315–5324 (2017)

37. Oliveira, S.A., Seguin, B., Kaplan, F.: dhSegment: a generic deep-learning approach for document segmentation. In: 2018 16th International Conference on Frontiers in Handwriting Recognition (ICFHR), pp. 7–12 (2018)

38. Wick, C., Puppe, F.: Fully convolutional neural networks for page segmentation of historical document images. In: 2018 13th IAPR International Workshop on Document Analysis Systems (DAS), pp. 287–292 (2018)

39. Chen, K., Seuret, M., Hennebert, J., Ingold, R.: Convolutional neural networks for page segmentation of historical document images. In: 2017 14th IAPR International Conference on Document Analysis and Recognition (ICDAR), vol. 1, pp. 965–970 (2017)

40. Mondal, R., Bhowmik, S., Sarkar, R.: tsegGAN: a generative adversarial network for segmenting touching nontext components from text ones in handwriting. IEEE Trans. Instrum. Meas. **70**, 1–10 (2020)

41. Khan, T., Mollah, A.F.: AUTNT-A component level dataset for text non-text classification and benchmarking with novel script invariant feature descriptors and D-CNN. Multimed. Tools Appl. **78**(22), 32159–32186 (2019)

42. Zhong, X., Tang, J., Yepes, A.J.: Publaynet: largest dataset ever for document layout analysis. In: 2019 International Conference on Document Analysis and Recognition (ICDAR), pp. 1015–1022 (2019)

43. Soto, C., Yoo, S.: Visual detection with context for document layout analysis. In: Proceedings of the 2019 Conference on Empirical Methods in Natural Language Processing and the 9th International Joint Conference on Natural Language Processing (EMNLP-IJCNLP), pp. 3464–3470 (2019)

44. Li, M., et al.: DocBank: a benchmark dataset for document layout analysis. arXiv Prepr. arXiv2006.01038 (2020)

45. Zhang, H., Xu, C., Shi, C., Bi, H., Li, Y., Mian, S.: HSCA-Net: a hybrid spatial-channel attention network in multiscale feature pyramid for document layout analysis. J. Artif. Intell. Technol. **3**(1), 10–17 (2023)

46. Embley, D.W., Hurst, M., Lopresti, D., Nagy, G.: Table-processing paradigms: a research survey. Int. J. Doc. Anal. Recognit. **8**, 66–86 (2006)

47. e Silva, A.C., Jorge, A.M., Torgo, L.: Design of an end-to-end method to extract information from tables. Int. J. Doc. Anal. Recognit. **8**(2–3), 144–171 (2006)

48. Yi, X., Gao, L., Liao, Y., Zhang, X., Liu, R., Jiang, Z.: CNN based page object detection in document images. In: 2017 14th IAPR International Conference on Document Analysis and Recognition (ICDAR), vol. 1, pp. 230–235 (2017)

49. Oliveira, D.A.B., Viana, M.P.: Fast CNN-based document layout analysis. In: Proceedings of the IEEE International Conference on Computer Vision Workshops, pp. 1173–1180 (2017)

50. Vo, N.D., Nguyen, K., Nguyen, T.V., Nguyen, K.: Ensemble of deep object detectors for page object detection. In: Proceedings of the 12th International Conference on Ubiquitous Information Management and Communication, pp. 1–6 (2018)

51. Huang, Y., et al.: A YOLO-based table detection method. In: 2019 International Conference on Document Analysis and Recognition (ICDAR), pp. 813–818 (2019)

52. Prasad, D., Gadpal, A., Kapadni, K., Visave, M., Sultanpure, K.: CascadeTabNet: an approach for end to end table detection and structure recognition from image-based documents. In: Proceedings of the IEEE/CVF Conference on Computer Vision and Pattern Recognition Workshops, pp. 572–573 (2020)

53. Siddiqui, S.A., Malik, M.I., Agne, S., Dengel, A., Ahmed, S.: Decnt: deep deformable cnn for table detection. IEEE Access. **6**, 74151–74161 (2018)

54. Agarwal, M., Mondal, A., Jawahar, C.V.: Cdec-net: composite deformable cascade network for table detection in document images. In: 2020 25th International Conference on Pattern Recognition (ICPR), pp. 9491–9498 (2021)
55. He, D., Cohen, S., Price, B., Kifer, D., Giles, C.L.: Multi-scale multi-task fcn for semantic page segmentation and table detection. In: 2017 14th IAPR International Conference on Document Analysis and Recognition (ICDAR), vol. 1, pp. 254–261 (2017)
56. Riba, P., Goldmann, L., Terrades, O.R., Rusticus, D., Fornés, A., Lladós, J.: Table detection in business document images by message passing networks. Pattern Recogn. **127**, 108641 (2022)
57. Schreiber, S., Agne, S., Wolf, I., Dengel, A., Ahmed, S.: Deepdesrt: deep learning for detection and structure recognition of tables in document images. In: 2017 14th IAPR International Conference on Document Analysis and Recognition (ICDAR), vol. 1, pp. 1162–1167 (2017)
58. Siddiqui, S.A., Khan, P.I., Dengel, A., Ahmed, S.: Rethinking semantic segmentation for table structure recognition in documents. In: 2019 International Conference on Document Analysis and Recognition (ICDAR), pp. 1397–1402 (2019)
59. Tensmeyer, C., Morariu, V.I., Price, B., Cohen, S., Martinez, T.: Deep splitting and merging for table structure decomposition. In: 2019 International Conference on Document Analysis and Recognition (ICDAR), pp. 114–121 (2019)
60. Khan, S.A., Khalid, S.M.D., Shahzad, M.A., Shafait, F.: Table structure extraction with bi-directional gated recurrent unit networks. In: 2019 International Conference on Document Analysis and Recognition (ICDAR), pp. 1366–1371 (2019)
61. Siddiqui, S.A., Fateh, I.A., Rizvi, S.T.R., Dengel, A., Ahmed, S.: Deeptabstr: deep learning based table structure recognition. In: 2019 International Conference on Document Analysis and Recognition (ICDAR), pp. 1403–1409 (2019)
62. Hashmi, K.A., Stricker, D., Liwicki, M., Afzal, M.N., Afzal, M.Z.: Guided table structure recognition through anchor optimization. IEEE Access. **9**, 113521–113534 (2021)
63. Deng, Y., Rosenberg, D., Mann, G.: Challenges in end-to-end neural scientific table recognition. In: 2019 International Conference on Document Analysis and Recognition (ICDAR), pp. 894–901 (2019)
64. Li, M., Cui, L., Huang, S., Wei, F., Zhou, M., Li, Z.: Tablebank: table benchmark for image-based table detection and recognition. In: Proceedings of the 12th Language Resources and Evaluation Conference, pp. 1918–1925 (2020)
65. Qasim, S.R., Mahmood, H., Shafait, F.: Rethinking table recognition using graph neural networks. In: 2019 International Conference on Document Analysis and Recognition (ICDAR), pp. 142–147 (2019)
66. Li, Y., Huang, Z., Yan, J., Zhou, Y., Ye, F., Liu, X.: GFTE: graph-based financial table extraction. In: Pattern Recognition. ICPR International Workshops and Challenges: Virtual Event, 10–15 Jan 2021, Proceedings, Part II, pp. 644–658 (2021)
67. Zheng, X., Burdick, D., Popa, L., Zhong, X., Wang, N.X.R.: Global table extractor (gte): a framework for joint table identification and cell structure recognition using visual context. In: Proceedings of the IEEE/CVF Winter Conference on Applications of Computer Vision, pp. 697–706 (2021)
68. Qiao, L., et al.: Lgpma: complicated table structure recognition with local and global pyramid mask alignment. In: Document Analysis and Recognition—ICDAR 2021: 16th International Conference, Lausanne, Switzerland, 5–10 Sept 2021, Proceedings, Part I, pp. 99–114 (2021)
69. Davila, K., Setlur, S., Doermann, D., Kota, B.U., Govindaraju, V.: Chart mining: a survey of methods for automated chart analysis. IEEE Trans. Pattern Anal. Mach. Intell. **43**(11), 3799–3819 (2020)
70. Svendsen, J.P.: Chart Detection and Recognition in Graphics Intensive Business Documents." (2015)
71. Siegel, N., Lourie, N., Power, R., Ammar, W.: Extracting scientific figures with distantly supervised neural networks. In: Proceedings of the 18th ACM/IEEE on Joint Conference on Digital Libraries, pp. 223–232 (2018)

72. Kavasidis, I., et al.: A saliency-based convolutional neural network for table and chart detection in digitized documents. In: Image Analysis and Processing—ICIAP 2019: 20th International Conference, Trento, Italy, 9–13 Sept 2019, Proceedings, Part II, vol. 20, pp. 292–302 (2019)

73. Li, P., Jiang, X., Shatkay, H.: Figure and caption extraction from biomedical documents. Bioinformatics. **35**(21), 4381–4388 (2019)

74. Siegel, N., Horvitz, Z., Levin, R., Divvala, S., Farhadi, A.: Figureseer: parsing result-figures in research papers. In: Computer Vision–ECCV 2016: 14th European Conference, Amsterdam, The Netherlands, 11–14 Oct 2016, Proceedings, Part VII, vol. 14, pp. 664–680 (2016)

75. Ray Choudhury, S., Mitra, P., Giles, C.L.: Automatic extraction of figures from scholarly documents. In: Proceedings of the 2015 ACM Symposium on Document Engineering, pp. 47–50 (2015)

76. Lopez, L.D., et al.: A framework for biomedical figure segmentation towards image-based document retrieval. BMC Syst. Biol. **7**, 1–16 (2013)

77. Clark, C., Divvala, S.: Pdffigures 2.0: mining figures from research papers. In: Proceedings of the 16th ACM/IEEE-CS on Joint Conference on Digital Libraries, pp. 143–152 (2016)

78. Praczyk, P.A., Nogueras-Iso, J.: Automatic extraction of figures from scientific publications in high-energy physics. Inf. Technol. Libr. **32**(4), 25–52 (2013)

79. Lee, P.-S., Howe, B.: Detecting and dismantling composite visualizations in the scientific literature. In: Pattern Recognition: Applications and Methods: 4th International Conference, ICPRAM 2015, Lisbon, Portugal, 10–12 Jan 2015, Revised Selected Papers, vol. 4, pp. 247–266 (2015)

80. Lee, P., West, J.D., Howe, B.: Viziometrics: analyzing visual information in the scientific literature. IEEE Trans. Big Data. **4**(1), 117–129 (2017)

81. Antani, S., Demner-Fushman, D., Li, J., Srinivasan, B.V., Thoma, G.R.: Exploring use of images in clinical articles for decision support in evidence-based medicine. In: Document Recognition and Retrieval XV, vol. 6815, pp. 230–239 (2008)

82. Zou, J., Thoma, G., Antani, S.: Unified deep neural network for segmentation and labeling of multipanel biomedical figures. J. Assoc. Inf. Sci. Technol. **71**(11), 1327–1340 (2020)

83. Zou, J., Antani, S., Thoma, G.: Localizing and recognizing labels for multi-panel figures in biomedical journals. In: 2017 14th IAPR International Conference on Document Analysis and Recognition (ICDAR), vol. 1, pp. 753–758 (2017)

84. Apostolova, E., You, D., Xue, Z., Antani, S., Demner-Fushman, D., Thoma, G.R.: Image retrieval from scientific publications: text and image content processing to separate multipanel figures. J. Am. Soc. Inf. Sci. Technol. **64**(5), 893–908 (2013)

85. Tsutsui, S., Crandall, D.J.: A data driven approach for compound figure separation using convolutional neural networks. In: 2017 14th IAPR International Conference on Document Analysis and Recognition (ICDAR), vol. 1, pp. 533–540 (2017)

86. Shi, X., Wu, Y., Cao, H., Burns, G., Natarajan, P.: Layout-aware subfigure decomposition for complex figures in the biomedical literature. In: ICASSP 2019–2019 IEEE International Conference on Acoustics, Speech and Signal Processing (ICASSP), pp. 1343–1347 (2019)

87. Cheng, B., Antani, S., Stanley, R.J., Thoma, G.R.: Automatic segmentation of subfigure image panels for multimodal biomedical document retrieval. In: Document Recognition and Retrieval XVIII, vol. 7874, pp. 294–304 (2011)

88. Nair, R.R., Sankaran, N., Nwogu, I., Govindaraju, V.: Automated analysis of line plots in documents. In: 2015 13th International Conference on Document Analysis and Recognition (ICDAR), pp. 796–800 (2015)

89. Al-Zaidy, R., Giles, C.: A machine learning approach for semantic structuring of scientific charts in scholarly documents. In: Proceedings of the AAAI Conference on Artificial Intelligence, vol. 31(2), pp. 4644–4649 (2017)

90. Ray Choudhury, S., Giles, C.L.: An architecture for information extraction from figures in digital libraries. In: Proceedings of the 24th International Conference on World Wide Web, pp. 667–672 (2015)

91. Zhou, Y.P., Tan, C.L.: Bar charts recognition using hough based syntactic segmentation. In: Theory and Application of Diagrams: First International Conference, Diagrams 2000 Edinburgh, Scotland, UK, 1–3 Sept 2000 Proceedings, vol. 1, pp. 494–497 (2000)
92. Zhou, Y.P., Tan, C.L.: Learning-based scientific chart recognition. In: 4th IAPR International Workshop on Graphics Recognition, GREC, vol. 7, pp. 482–492 (2001)
93. Shi, Y., Wei, Y., Wu, T., Liu, Q.: Statistical graph classification in intelligent mathematics problem solving system for high school student. In: 2017 12th International Conference on Computer Science and Education (ICCSE), pp. 645–650 (2017)
94. Choudhury, S.R., Wang, S., Mitra, P., Giles, C.L.: Automated data extraction from scholarly line graphs. In: GREC, Nancy, France (2015)
95. Huang, W., Zong, S., Tan, C.L.: Chart image classification using multiple-instance learning. In: 2007 IEEE Workshop on Applications of Computer Vision (WACV'07), p. 27 (2007)
96. Gao, J., Zhou, Y., Barner, K.E.: View: visual information extraction widget for improving chart images accessibility. In: 2012 19th IEEE International Conference on Image Processing, pp. 2865–2868 (2012)
97. Chagas, P., et al.: Evaluation of convolutional neural network architectures for chart image classification. In: 2018 International Joint Conference on Neural Networks (IJCNN), pp. 1–8 (2018)
98. Poco, J., Heer, J.: Reverse-engineering visualizations: recovering visual encodings from chart images. Comput. Graph. Forum. **36**(3), 353–363 (2017)
99. Kaur, P., Kiesel, D.: Combining image and caption analysis for classifying charts in biodiversity texts. In: VISIGRAPP (3: IVAPP), pp. 157–168 (2020)
100. Huang, S.: An Image Classification Tool of Wikimedia Commons. Humboldt-Universität zu Berlin (2020)
101. Bajić, F., Job, J., Nenadić, K.: Data visualization classification using simple convolutional neural network model. Int. J. Electr. Comput. Eng. Syst. **11**(1), 43–51 (2020)
102. Balaji, A., Ramanathan, T., Sonathi, V.: Chart-Text: A Fully Automated Chart Image Descriptor. arXiv Prepr. arXiv1812.10636 (2018)
103. Araújo, T., Chagas, P., Alves, J., Santos, C., Sousa Santos, B., Serique Meiguins, B.: A real-world approach on the problem of chart recognition using classification, detection and perspective correction. Sensors. **20**(16), 4370 (2020)
104. Dai, W., Wang, M., Niu, Z., Zhang, J.: Chart decoder: generating textual and numeric information from chart images automatically. J. Vis. Lang. Comput. **48**, 101–109 (2018)
105. Choi, J., Jung, S., Park, D.G., Choo, J., Elmqvist, N.: Visualizing for the non-visual: enabling the visually impaired to use visualization. Comput. Graph. Forum. **38**(3), 249–260 (2019)
106. Jung, D., et al.: Chartsense: interactive data extraction from chart images. In: Proceedings of the 2017 CHI Conference on Human Factors in Computing Systems, pp. 6706–6717 (2017)
107. Thiyam, J., Singh, S.R., Bora, P.K.: Challenges in chart image classification: a comparative study of different deep learning methods. In: Proceedings of the 21st ACM Symposium on Document Engineering, pp. 1–4 (2021)
108. Yang, L., Huang, W., Tan, C.L.: Semi-automatic ground truth generation for chart image recognition. In: Document Analysis Systems VII: 7th International Workshop, DAS 2006, Nelson, New Zealand, 13–15 Feb 2006. Proceedings, vol. 7, pp. 324–335 (2006)
109. Weihua, H.: Scientific Chart Image Recognition and Interpretation (2008)
110. Obeid, J., Hoque, E.: Chart-to-Text: Generating Natural Language Descriptions for Charts by Adapting the Transformer Model. arXiv Prepr. arXiv2010.09142 (2020)
111. Chen, L., Zhao, K.: An approach for chart description generation in cyber–physical–social system. Symmetry (Basel). **13**(9), 1552 (2021)
112. Liu, X., Klabjan, D., Bless, P.N.: Data Extraction from Charts via Single Deep Neural Network. arXiv Prepr. arXiv1906.11906 (2019)
113. Dadhich, K., Daggubati, S.C., Sreevalsan-Nair, J.: BarChartAnalyzer: digitizing images of bar charts. In: IMPROVE, pp. 17–28 (2021)

114. Sohn, C., Choi, H., Kim, K., Park, J., Noh, J.: Line chart understanding with convolutional neural network. Electronics. **10**(6), 749 (2021)
115. Demir, S., Carberry, S., McCoy, K.F.: Summarizing information graphics textually. Comput. Linguist. **38**(3), 527–574 (2012)
116. Al-Zaidy, R.A., Choudhury, S.R., Giles, C.L.: Automatic summary generation for scientific data charts. In: 30th AAAI Conference on Artificial Intelligence, AAAI 2016, pp. 658–663 (2016)
117. UCI Machine Learning Repository. http://archive.ics.uci.edu/ml/datasets/Newspaper+and +magazine+images+segmentation+dataset#
118. The MediaTeam Document Database II. http://www.mediateam.oulu.fi/downloads/MTDB/
119. Clausner, S.P.C., Antonacopoulos, A.: ICDAR2019 competition on recognition of documents with complex layouts—RDCL2019. In: Proceedings of the 15th International Conference on Document Analysis and Recognition (ICDAR2019), pp. 1521–1526 (2019)
120. Clausner, C., Antonacopoulos, A., Pletschacher, S.: ICDAR2017 competition on recognition of documents with complex layouts—RDCL2017. In: Proceedings of the International Conference on Document Analysis and Recognition, ICDAR, vol. 1, pp. 1404–1410 (2017). https://doi.org/10.1109/ICDAR.2017.229
121. Antonacopoulos, A., Bridson, D., Papadopoulos, C., Pletschacher, S.: A realistic dataset for performance evaluation of document layout analysis. In: 10th International Conference on Document Analysis and Recognition, 2009. ICDAR'09, pp. 296–300 (2009)
122. UW-III English/Technical Document Image Database. http://isis-data.science.uva.nl/events/ dlia//datasets/uwash3.html
123. Indermühle, E., Liwicki, M., Bunke, H.: IAMonDo-database: an online handwritten document database with non-uniform contents. In: Proceedings of the 9th IAPR International Workshop on Document Analysis Systems, pp. 97–104 (2010)
124. Matsushita, T., Nakagawa, M.: A database of on-line handwritten mixed objects named "Kondate". In: 2014 14th International Conference on Frontiers in Handwriting Recognition (ICFHR), pp. 369–374 (2014)
125. Zhou, X.-D., Liu, C.-L.: Text/non-text ink stroke classification in Japanese handwriting based on Markov random fields. In: Ninth International Conference on Document Analysis and Recognition, 2007. ICDAR 2007, vol. 1, pp. 377–381 (2007)
126. Delaye, A., Liu, C.-L.: Contextual text/non-text stroke classification in online handwritten notes with conditional random fields. Pattern Recogn. **47**(3), 959–968 (2014)
127. School of Cultural Texts and Records. Bichitra: Online Tagore Variorum. http://bichitra.jdvu. ac.in/index.php. Accessed 6 Nov 2017
128. Islamic Heritage Project (IHP) collection. http://ocp.hul.harvard.edu/ihp/
129. Gao, L., et al.: ICDAR 2019 competition on table detection and recognition (cTDaR). In: 2019 International Conference on Document Analysis and Recognition (ICDAR), pp. 1510–1515 (2019)
130. Mondal, A., Lipps, P., Jawahar, C.V.: IIIT-AR-13K: a new dataset for graphical object detection in documents. In: Document Analysis Systems: 14th IAPR International Workshop, DAS 2020, Wuhan, China, 26–29 Jul 2020, Proceedings, vol. 14, pp. 216–230 (2020)
131. Zhong, X., ShafieiBavani, E., Jimeno Yepes, A.: Image-based table recognition: data, model, and evaluation. In: Computer Vision–ECCV 2020: 16th European Conference, Glasgow, UK, 23–28 Aug 2020, Proceedings, Part XXI, vol. 16, pp. 564–580 (2020)
132. Lee, S.L., Zare, M.R.: Biomedical compound figure detection using deep learning and fusion techniques. IET Image Process. **12**(6), 1031–1037 (2018)
133. Zhang, J., Xie, Y., Wu, Q., Xia, Y.: Medical image classification using synergic deep learning. Med. Image Anal. **54**, 10–19 (2019)
134. Jobin, K.V., Mondal, A., Jawahar, C.V.: Docfigure: a dataset for scientific document figure classification. In: 2019 International Conference on Document Analysis and Recognition Workshops (ICDARW), vol. 1, pp. 74–79 (2019)

135. Tang, B., et al.: Deepchart: combining deep convolutional networks and deep belief networks in chart classification. Signal Process. **124**, 156–161 (2016)
136. Davila, K., et al.: ICDAR 2019 competition on harvesting raw tables from infographics (chart-infographics). In: 2019 International Conference on Document Analysis and Recognition (ICDAR), pp. 1594–1599 (2019)
137. Greenbacker, C., Wu, P., Carberry, S., McCoy, K.F., Elzer, S.: Abstractive summarization of line graphs from popular media. In: Proceedings of the Workshop on Automatic Summarization for Different Genres, Media, and Languages, pp. 41–48 (2011)
138. Burns, R., Carberry, S., Elzer Schwartz, S.: An automated approach for the recognition of intended messages in grouped bar charts. Comput. Intell. **35**(4), 955–1002 (2019)
139. Böschen, F., Beck, T., Scherp, A.: Survey and empirical comparison of different approaches for text extraction from scholarly figures. Multimed. Tools Appl. **77**, 29475–29505 (2018)
140. Kafle, K., Price, B., Cohen, S., Kanan, C.: Dvqa: understanding data visualizations via question answering. In: Proceedings of the IEEE Conference on Computer Vision and Pattern Recognition, pp. 5648–5656 (2018)
141. Fletcher, L.A., Kasturi, R.: A robust algorithm for text string separation from mixed text/ graphics images. Pattern Anal. Mach. Intell. IEEE Trans. **10**(6), 910–918 (1988)
142. Göbel, M., Hassan, T., Oro, E., Orsi, G.: ICDAR 2013 table competition. In: 2013 12th International Conference on Document Analysis and Recognition, pp. 1449–1453 (2013)
143. Raja, S., Mondal, A., Jawahar, C.V.: Table structure recognition using top-down and bottom-up cues. In: Computer Vision–ECCV 2020: 16th European Conference, Glasgow, UK, 23–28 Aug 2020, Proceedings, Part XXVIII, vol. 16, pp. 70–86 (2020)
144. Taschwer, M., Marques, O.: Automatic separation of compound figures in scientific articles. Multimed. Tools Appl. **77**, 519–548 (2018)
145. Zhang, R., et al.: ICDAR 2019 robust reading challenge on reading chinese text on signboard. In: 2019 International Conference on Document Analysis and Recognition (ICDAR), pp. 1577–1581 (2019)
146. Zhu, J., Ran, J., Lee, R.K., Choo, K., Li, Z.: AutoChart: A Dataset for Chart-To-Text Generation Task. arXiv Prepr. arXiv2108.06897 (2021)
147. Sai, A.B., Mohankumar, A.K., Khapra, M.M.: A survey of evaluation metrics used for NLG systems. ACM Comput. Surv. **55**(2), 1–39 (2022)

Chapter 5
Case Study

Abstract Historical documents carry many crucial information regarding a place or person for a specific time period, whereas contemporary documents cover recent status. However, contemporary documents possess mostly non-Manhattan layout, even overlapping layout in some cases. In contrast to that layouts of historical documents are comparably constrained, but these documents suffer from huge degradations. These facts say analysis of layout present in documents from these categories are very challenging. In this chapter, a method named as ABCD (Analysis of Basic Contents in Documents) is presented, which automatically segments and classifies the regions of early Indian printed document images. This method was submitted to *REID2019* competition for Task 1, and it outperforms all other participating methods from different industrial and academic sectors in terms of layout analysis and becomes the winner. In this chapter, also a method name BINYAS is presented which was developed to deal with contemporary documents. This method secures the 3rd position in *RDCL2019* competition.

Keywords Segmentation · Distance transform · Morphology · Historical document · Text non-text · REID · ABCD · BINYAS · RDCL

5.1 Analysis of Basic Contents in Documents (ABCD)

ABCD is developed for the layout analysis of historical printed documents. Initially, the input document image is binarized and pre-processed. Then a connected component analysis is performed on the binarized image to suppress the non-text components and noise. This step generates one text and one non-text-only image. Further on the text image, an iterative, and adaptive morphology-based operation is performed to generate the segmented text regions. Lastly, the segmented text and non-text regions are combined to generate the final segmented image. A diagrammatic representation of the work is given in Fig. 5.1.

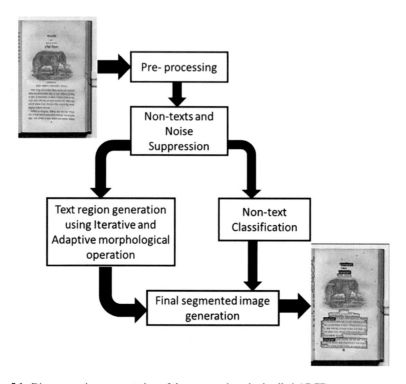

Fig. 5.1 Diagrammatic representation of the proposed method called ABCD

5.1.1   Pre-processing

In this step, an input document image I_{input} is initially binarized using a local threshold based method [1]. Historical document images often possess huge intensity variation due to noise. Therefore, a local threshold based binarization may produce many tiny discrete components. However, these can also appear when an actual component gets fragmented due to intensity level variation. This, in turn, produces the wrong boundary of the original component (see Fig. 5.2a). So, to preserve the actual boundary of original components and to merge the fragmented tiny components to their source components, distance transformation on the binarized image I_{bin} is performed.

Distance transformation is a morphological operation, which replaces each data pixel in a binarized image with the distance of it from the nearest boundary.

Let S is a set of data pixels and for the location of each ith data pixel (x_i, y_i) in S, distance to the nearest boundary point location (x_j, y_j) in S can be computed as

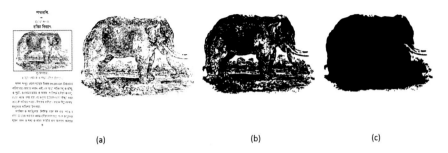

(a)   (b)   (c)

Fig. 5.2 Represents (**a**) a fragmented non-text component with many tiny clippings, cropped from a binarized image (**b**) the same non-text component after applying distance transformation, and (**c**) the non-text component in (**b**) after component wise morphological region filling

$$DT_S((x_i, y_i)) = \min\nolimits_{((x_j, y_j) \in S)} \left\{ \sqrt{(x_i - x_j)^2 + (y_i - y_j)^2} \right\} \qquad (5.1)$$

In Eq. (5.1), the Euclidean distance is considered as the distance measure. While performing the distance transformation, the input binary image is inverted, so that the background pixels become the data pixels and the closest background pixels to the component boundary can be identified. This is because, few of these background pixels appear in between the broken parts of the component boundary and the objective here is to convert these background pixels to the data pixels to seal the leak. To attain that thresholding is performed on the transformed image with a significantly small threshold value. This process stiches the broken boundary of the components and generates the thresholded image I_{th} (see Fig. 5.2b).

Now to join the fragmented clippings present inside a component, a component wise morphological region filling operation on I_{th} is performed (see Fig. 5.2c). However, in this process, CCs which have height and width less than 50% of the original page height and width are only considered. This is because to avoid the margins and margin like components.

5.1.2   Non-text Suppression and Noise Removal

In historical printed books, large images, margins, straight lines, seals, and different page decorative are commonly found which are termed as non-text components. Usually, these components are larger than the body text components either in terms of height or width or both. Therefore, the dimensions of a component can play an important role to separate out the non-text CCs from the text CCs.

Keeping the above fact in mind, the CCs from I_{th} are initially extracted and then the histograms of height and width of the components $Hist_h$ and $Hist_w$ respectively are computed. As said earlier, in historical documents with huge background variation, binarization may cause some random noise appear as tiny components.

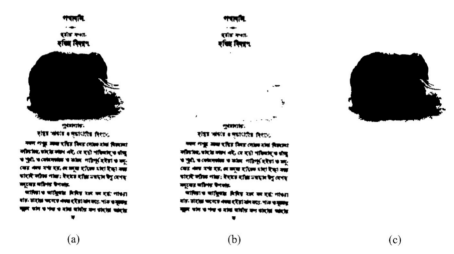

Fig. 5.3 Shows (a) the pre-processed image I_{th}, and the corresponding (b) text-only image I_t and (c) non-text only image I_{nt}

However, the number of constituting foreground pixels in these components is very less in comparison to other CCs. Therefore, before computing those histograms, CCs with less than six constituent data pixels in it are filtered out.

Now, from $Hist_h$ and $Hist_w$, the three most occurring heights and widths are selected separately. Let $\{f_1^h, f_2^h, f_3^h\}$ and $\{f_1^w, f_2^w, f_3^w\}$ are the sets of three most occurring heights and widths respectively. From these, $th^h = \dfrac{\sum_{i=1}^{3} f_i^h}{3}$ and $th^w = \dfrac{\sum_{i=1}^{3} f_i^w}{3}$ are computed as the threshold values for height and width to identify the large components. Based on these, the text-only image I_t and non-text only image I_{nt} are generated (see Fig. 5.3) as follows,

$$
\begin{cases}
CC_i \in I_{nt} & \text{if } h_i > th^h \text{ or } w_i > th^w \\
CC_i \in I_t & \qquad\quad otherwise
\end{cases}
\tag{5.2}
$$

Here, h_i and w_i represent the height and width of CC_i respectively.

5.1.3   Text Region Generation

In this step, the text regions are initially generated and then refined to get paragraphs. Therefore, in this section, first, the region generation process is explained and then the region refinement as follows

Region Generation

For the generation of text regions, an iterative and adaptive morphological region growing (IAMRG) operation is followed. For that purpose, a rectangular structuring element is considered. In each iteration, the height and width of this structuring element are set dynamically. Additionally, in each iteration, the said structuring element is applied twice one with no rotation and again with $90°$ rotation to grow the merged region both horizontally and vertically. At each iteration of the region merging process, morphological closing is performed with a rotating structuring element whose dimensions are set dynamically. The region generation steps are given in Algorithm 5.1.

In the algorithm, γ is the difference factor and the value of it is experimentally set as 0.2. Here α is the control parameter, which prevents the premature convergence of the region growing process. That means, as in each iteration of IAMRG, the components of I_t are merged and the dimensions of the structuring element $SE_{(h,w)}$ are set accordingly, in certain cases, it is observed that the dimensions of $SE_{(h,w)}$ grow rapidly. As a result, text components from different columns may get merged. Thus, it may produce a single region even for the pages with two columns. So, to limit the growth of $SE_{(h,w)}$ in size, the parameter α is introduced as control parameter. Here the value of α is set as 2.

Algorithm 5.1

$IAMRG(I_t, \gamma, \alpha)$

Let the image I_t has n_i number of text components at the ith iteration of the region generation process and $SE_{(h,w)}$ represents a rectangular structuring element with no rotation, where h and w represent the height and width of the structuring element respectively. Suppose H_{cc} and W_{cc} represent the height and width histograms of the components present in I_t respectively.

1. **WHILE** $(n_i > \gamma \times n_{i-1})$
   1.1 Compute H_{cc} and W_{cc}.
   1.2 Compute the dimensions of $SE_{(h,w)}$ as
   1.2.1 $h = argmax\{H_{cc}\}/\alpha$
   1.2.2 $w = argmax\{W_{cc}\}/\alpha$
   1.3 Perform the following morphological operation on I_t
   1.3.1 $I_t = ((((I_t \oplus SE_{(h,w)}) \ominus SE_{(h,w)}) \oplus SE_{(w,h)}) \ominus SE_{(w,h)})$
2. **END**

Region Refinement

The purpose of this step is to refine the generated text regions so that the region boundaries form the closest polygon. In addition to that, in this step, the paragraph cut points are also identified to perform the paragraph level segmentation (see Fig. 5.4c). The entire region refinement process is given in Algorithm 5.2.

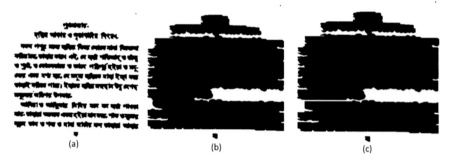

Fig. 5.4 Represents (**a**) text-only image I_t, (**b**) after text region generation and (**c**) after region refinement and paragraph identification

In Algorithm 5.2, μ is threshold defined in terms of row-wise data pixel count, which is experimentally set as 2. δ is the left white space threshold, i.e. when a left side white space exceeds this value, it is considered as left indent. Left indents are mostly used to indicate the starting of a paragraph. As a common practice, in most of the printed materials, the length of left indents is kept between 0.17 and 0.50 in. That is around 16 to 50 in terms of number of pixels. So, the value of δ is set as 30. β is the right white space threshold to indicate right indents. Right indents are often useful to get an idea about the end of a paragraph. So, for that the value of β is set to 150.

Algorithm 5.2

REFINE(μ, δ, β)

1 *For each CC in the segmented image*

   1.1 *Extract the corresponding text region from I_t*

   1.2 *Scan each row of the extracted text region to count the number of text pixels (here '0') i.e. P*

   1.3 *If P > μ*

      1.3.1 *Fill the row with '0' which generates a black strip for each text line*

      1.3.2 *Prune each black strip from left and right sides according to the left most and the right most text pixels present in the corresponding text line.*

      1.3.3 *Compute the left and right white space profiles which appear at the left and right sides of the pruned black strips respectively.*

      1.3.4 *Among these left white spaces, the ones with more than δ number of background pixels present in a row are considered as left-paragraph-cut-point (LPCP).*

      Among these right white spaces, the ones with more than β number of background pixels present in a row are considered as right-paragraph-cut-point (RPCP).

(continued)

Algorithm 5.2 (continued)
    1.3.5 *Consider two consecutive pruned black strips and perform the following operations:*
    1.3.5.1 *If (out of two consecutive pruned black strips, the 1st one have the RPCP or the 2nd one have the LPCP)*
    1.3.5.1.1 *Do nothing*
    1.3.5.2 *Else*
    1.3.5.2.1 *Fill the white strip between the pruned black strips under consideration to merge them.*
    1.3.5.3 *End*
    1.4 *End*
  2 *End*

5.1.4 Non-text Classification

Any document may have margins, separators, tables, images and seals, which are considered as non-text components. In the present scope of the work these components are recognized based on methods described in the following sub-sections.

Identification of Separator and Margin

Such CCs are mostly straight line based. However, in some historical documents decorative margins are often found. On the other hand, separators are either horizontally or vertically aligned straight lines. In this work, both of these components are considered as separator.

To recognize these CCs, in addition to height and width, the aspect ratio, i.e. acc_i is also computed and the solidity sol_i of each CC_i is obtained from I_{nt} as

$$acc_i = \min\{h_i, w_i\}/\max\{h_i, w_i\} \tag{5.3}$$

$$sol_i = \textit{Number of data pixels}/(h_i \times w_i) \tag{5.4}$$

Now, a CC_i is recognized as separator if and only if

$$\{(acc_i \geq 0.2) \wedge (\max\{h_i, w_i\} \geq 100) \wedge (sol_i \geq 0.8)\} \\ \vee \{((h_i \geq 0.7 \times H) \wedge ((w_i \geq 0.7 \times W) \wedge (sol_i \leq 0.02)\} \tag{5.5}$$

In Eq. (5.5), H and W represent the height and width of the input image. All the thresholds are computed experimentally on the training set.

The threshold on height and width in the first part of Eq. (5.5) is to avoid those text CCs which share the aspect ratio similar to line separators. For example, 'i', 'l' in English. But these line separators often possess high solidity. That is why the threshold on the solidity at the first part of Eq. (5.5) is set as 0.8. On the other hand, margins are large in height and width but have less solidity. Therefore, the thresholds at the second part of Eq. (5.5) are set so.

Identification of Table

Tables can be fully connected, partially connected or completely disjoint. However, in the present scope of the work, fully connected tables are only considered.

Fully connected tables have closed cells. In this work, these cells are extracted as separate component, which can be said as *cell component.* To do that, the components are inverted so that the regions inside them can be extracted as components. To identify a CC as table, the number of *cell component* and the sum of the height, width and area of these cell components are computed.

Let CC_count_i is the number of cell components in a given CC_i and ch_{sum}, cw_{sum}, and A_{sum} represent the sum of heights, widths and areas of these cell components respectively. If A_i represents the area of the given CC_i, then it is recognized as table if and only if

$$CC\_count_i \geq 4 \wedge (ch_{sum}, \geq 0.75 \times h_i) \wedge (cw_{sum} \geq 0.75 \times w_i) \wedge (A_{sum} \times A_i) \quad (5.6)$$

In Eq. (5.6), all the threshold values are computed experimentally on the training set. Rest of the components in I_{nt}, which are not recognized as separator or table are classified as image.

5.1.5 *Experimental Results*

The proposed method is evaluated using the database of *"ICDAR 2019 Competition on Recognition of Early Indian Printed Documents—REID2019"* [2] This database contains 81 document images, taken from the digitized pages of the early (1731–1914) Indian printed books written in Bengali. These books have been digitized by the British Library as a part of their on-going digitization project. The pages in this database have several issues related to layout and OCR like *decoration, non-straight lines, non-rectangular shaped regions, presence of separator, varying font size, varying column width*, etc. Additionally, as historical documents, these possess many quality level degradations like *bleed through, fade ink* etc. degradations due to aging and scan related issues are also there.

For the evaluation, out of these 81 pages 25 pages are made available by the REID2019 organizers with ground truth for training purpose and rest 56 pages are used for testing. REID2019 had two challenges defined as Task: (a) recognition or segmentation of the pages, and (b) recognition of Multi Lingual Tabular Data. But all the participating methods had appeared for Task 1 only. The present method too was

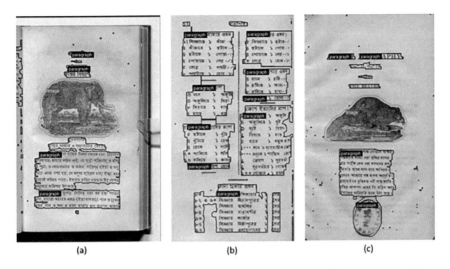

(a)   (b)   (c)

Fig. 5.5 Displays the page segmentation and classification outputs of the present method on samples taken from REID 2019 dataset. Here the regions with blue outlines are texts, with magenta outlines are separators, with mint outlines are images

Table 5.1 Comparative assessment between the proposed method and other participating methods of *REID 2019*

Method	Segmentation + classification (in %)	Text region only (in %)
BanglaOCR	58.5	64.4
DS	58.9	62.8
Google multi-lingual OCR	67.9	**80.4**
Tesseract 4.0	67.4	72.6
ABCD (proposed method)	**69.9**	74.2

submitted for Task 1 only and it was declared as the winning method for the Task 1 of REID 2019. Few outputs of the proposed method are given in Fig. 5.5.

The layout or segmentation performances are evaluated from two aspects; one by comparing the regions generated by a method with ground truth regions and other by quantifying different types of errors like miss/partial miss, merge, split, and false detection [3]. In case of classification, region types are also compared. Each error type is quantified by their significance. For the significance of error different evaluation profiles are used. A detail result by comparing the performance of the present method with other participating methods, i.e. *BanglaOCR, DS, Google Multi-Lingual OCR* and *Tesseract 4.0* is given in Table 5.1.

From Table 5.1, it can be observed that the proposed method outperforms the other participating methods in segmentation and region classification task (i.e. Task 1). But in case of text only regions it secures the second position, mostly because in the present scope of the work the seals are not considered as non-text. Therefore, most of the seals are recognized as text (see Fig. 5.5c). On the other hand, in this

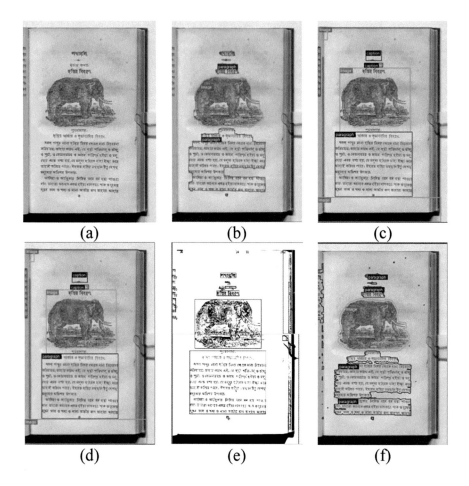

Fig. 5.6 Presents (**a**) the original image, (**b**) the ground truth image, (**c**) the output of *Tesseract 4.0*, (**d**) the output of *Tesseract 3.0*, (**e**) the output of *Recogniform Layout Analysis SDK*, and (**f**) the output of *ABCD* (**proposed work**). Here, in (**b**, **c**, **d** and **f**), regions with blue outlines are texts, with magenta outlines are separators, with mint outlines are images. In (e), blue indicates text region, red indicates non-text, yellow indicates line, and gray indicates noise

work, a very simple pre-processing technique is followed. Due to that many noise are mistakenly detected as text.

In addition to the numerical results, an image level comparison is presented among the outputs of few freely available software products and ABCD in Fig. 5.6. These software products include *Tesseract 4.0* [4], *Tesseract 3.0* and *Recogniform Layout Analysis SDK* [5].

5.2 BINYAS

BINYAS is a DLA method developed for contemporary documents. This can automatically segment an input document image in its constituent regions and classify these into text and non-text regions. The text regions are further classified as heading, drop-capital, header, footer, and paragraph. The non-texts are also classified as separator, image, bar chart, table, and inverted text. BINYAS follows a hybrid approach to segment the input image and is designed mostly to handle complex layouts. In this section a summarized view of the DLA method BINYAS is presented, more detailed discussion and performance analysis can be found in [6].

5.2.1 Pre-processing

Recent methods that consider binarized image as input, either use global or local threshold based binarization method at the pre-processing stage. However, a global threshold may cause in attaching two or more body text components together due to intensity level variation of the background. For example, if the intensity of foreground and background is almost similar within a text region then such situation may occur. Touching of body text components may create problem in the subsequent steps. On the other hand, a local threshold based binarization technique generates multiple small components for a single CC; thereby fails to generate a clear outer boundary. To overcome this problem, BINYAS considers both the globally and locally thresholded binarized images. The combined use of these images not only helps to get a clear outer boundary for the non-texts but also prevents body texts to be attached with each other.

For the generation of globally thresholded binarized image, first contrast stretching operation is performed on the input images so that a clear intensity difference can be achieved between the foreground and the background, and then binarized using a global thresholding based method. For the generation of locally thresholded binarized image, the window size is estimated dynamically based on the most frequent height and width of the components present in the input image. Once the window size is estimated, the input image is binarized locally using it.

5.2.2 Isolation of Separators

Separators mostly include straight line like components. However, some connected horizontal and vertical lines generally called *margins* can also appear as separators. Separators play a crucial role in defining the structure of a document so these should be identified carefully. In the present work, we try to capture all possible types of separators by classifying the extracted CCs based on some common properties of these. For example, the straight line like components generally have higher solidity with significant difference between their height and width. In addition to that, margins acquire major area in a document image with less solidity.

However, some text components may also share the similar nature of shapes. For example, consider 'i', 'l', etc. in Roman script. Hence, only these properties are not enough to identify those separators. Thus, some thresholds are considered on height and width to avoid text components with similar shape.

5.2.3   Layout Complexity Estimation

In BINYAS, it is identified whether the input image is geometrically simple or complex from the layout point of view, so that the intricate operations, designed mainly to deal with images having complex layout can be avoided, while processing the simple documents. This not only reduces the processing time but also helps to avoid the unwanted errors that may creep in due to the application of such operations on simple documents. For the complexity estimation, the components height variation and a threshold on the most frequently occurred component height are considered.

5.2.4   Separation of Large and Small Components

This operation is performed only for the documents with geometrically complex layout. BINYAS follows a morphology based approach for region generation and for any such method, the dimension of the structuring element is a matter of concern. Estimation of this dimension becomes more complex when an input image possesses a wide variation of components in terms of size. To deal with this, in this work, the components of a geometrically complex document image are partitioned into two different images; an image with large components and an image with small components by setting a threshold on the component height. This in turn makes the computation easier as in each partition the size variation of components becomes manageable.

5.2.5   Text and Non-text Separation

The presence of arbitrary shaped non-texts like graphics, images, charts, tables, and others not only makes the contemporary documents more illustrative and decorative but also challenging in terms of layout analysis. Additionally, we cannot perform OCR on these components. For that reason, non-texts should be identified and separated from the texts, so that these can be processed independently. In this work, text non-text separation is done by performing component analysis independently on the image with large components and image with small components.

The image with large components mostly contains title texts, drop capitals (if present) as text that need to be separated from the non-text components. To serve the purpose, following information of a component are captured.

- *Degree of alignment:* This feature estimates the number of left and right neighbors of a component. But to consider a component as the neighbor of the component under consideration, a concept called *region of horizontal neighborhood* (ROHN) is introduced. Every component has its own ROHN, when the ROHN of one component overlaps the ROHN of another with a certain percentage then only these are considered as the neighbor of each other. In this way, the left and right neighboring components are identified for a particular component.
- *Number of closed region:* Closed regions inside a component imply the hollow regions appear due to the intrinsic shape structure of the component or sometimes due to the intensity variation in formation of it. In document images, non-text components like images, graphics that are used for decoration generally have high intensity level variation compares to the text, thus binarization of these components sometime generates a huge number of hollow regions within these. On the other hand, though closed regions generated inside the text components mostly due to the shape structure of the component but they are less in number. Thus, here, the said feature is used to identify the non-text.
- *Number of elementary components:* This feature helps in finding the non-text, as mostly a single text component appears as a single component, but due to intensity difference after binarization a non-text component may get broken into many elementary components.

In an image with small components, most of the components are body texts apart from few non-text components such as bullets and broken parts of large non-text components. During binarization, some tiny parts may get separated from a large non-text component due to its intensity level variation, and these appear as non-text in the said image. As these tiny components possess some clearly distinct nature in comparison to text components, hence a simplistic approach is sufficient to separate them from the texts. So, some simple properties of the components like frequent component height, number of closed regions and solidity are considered for that purpose.

5.2.6 Thickness Based Text Separation

Thickness based text separation is performed on the images with small text components to separate the bold texts from the normal body texts. These bold words or sentences serve some special purposes in document images such as in newspapers these may act as headlines. Thus, these should be enclosed in separate regions to distinguish from the normal body texts. Therefore, BINYAS estimates the thickness of a component. For that a weighted sum of horizontal average run-lengths and vertical average run-lengths is computed. While adjusting the weights the height-major and width-major nature of the components are considered. These weights are set dynamically for each component according to their shape nature.

5.2.7   Text Region Segmentation

The text region segmentation is performed on the text images only. In the region generation process, the objective is to merge the text components that lie very closely with each other within a document image. To do that, in this work, an adaptive morphological operation is iteratively performed on the components of an input image. It is to be noted that the performance of any morphology based method heavily depends on the design and use of structuring element. Though in this work, a rectangular structuring element is used, but the dimensions of the same are set dynamically which get updated after every iteration. Additionally, to serve the purpose, at each iteration the structuring element is applied twice, first with no rotation and second with 90° rotation of the same.

Most of the morphology based methods either use fixed sized structuring element or perform one-time dimension estimation. But in this work, the dimension of the structuring element set dynamically at each iteration. Additionally, none of the recent methods uses rotating structuring element like the present one, which helps in merging not only two horizontally neighboring components but also two vertically neighboring components in at one go. Thus, this approach takes a smaller number of iterations to merge the text components for generating the region.

The regions are refined to get the closest region boundary and further partitioned to generate paragraphs. For paragraph segmentation, the text indentation rules which commonly followed in general documents are used.

5.2.8   Non-text Classification

Similar to the texts, non-texts also play an important role in documents. Most of these components are used to illustrate different sort of information. But to extract such information, these components need to be processed differently according to their types. Thus, classification of non-texts becomes an integral part of a comprehensive DLA system. In this work, the non-texts are classified as bar chart, table and image. Inverted texts either classified as separators or as non-texts.

For bar-chart identification, in the present invention, the horizontal projection profile (HPP) of the components is computed. It is to be noted that a component presenting bar-chart always has some clear peaks for the bars and troughs for the intermediate regions. In the present work, any local maxima is considered as a valid peak if it is greater than its two immediate neighboring local minima by a threshold.

The value of the threshold is kept as 30, based on the estimated frequent component height. If the number of such peaks is greater than 2, then the component is classified as bar chart. In some cases, bars in bar-chart do not appear as a single component. In that case the rectangular shaped components are identified based on aspect ratio and solidity of the component and components with almost same lower base are connected to generate the bar-chart.

After the identification of bar-charts, rest of the non-text components are examined to check whether any inverted texts are there or not. For this, both the HPP and vertical projection profile (VPP) of a component are computed. Due to the inter character spacing, there may be many peaks in the HPP of a component representing an inverted text. Whereas if the inverted text component contains single line then its VPP may have a single and clear trough region, and if it contains multiple lines then there will be multiple clear trough regions. Considering this fact, in this work, whether the component height is greater than a threshold is checked first, and if so how many peaks are there in its VPP and HPP. If it is more than 4 then it is identified as multi-line inverted text. On the other hand, if component's height is less than the threshold and the number of peaks in VPP is 1 and HPP is more than 4, then it is considered as single line inverted text.

Using bounding box inspection, the rectangular shaped boxes are extracted. It is assumed that in a table, it will have proper rows and columns. The bounding box regions are cropped from the original image, and are converted to grayscale and then binarized. Now the obtained binary image is processed. CCs are again extracted to retrieve each cell of the table. The span of all extracted cells is then calculated. A cell is the intersection of a row and a column. As it is expected that all the cells together should form the span of the whole component, if combined span of all extracted cell is greater than 75% of the span of the original component, then it is considered as table.

References

1. Bhowmik, S., Sarkar, R., Das, B., Doermann, D.: GiB: a game theory inspired binarization technique for degraded document images. IEEE Trans. Image Process. **28**(3), 1443–1455 (2019)
2. Clausner, C., Antonacopoulos, A., Derrick, T., Pletschacher, S.: ICDAR2019 competition on recognition of early Indian printed documents—REID2019. In: 2019 International Conference on Document Analysis and Recognition (ICDAR), pp. 1527–1532 (2019)
3. Pletschacher, S., Antonacopoulos, A.: The PAGE (page analysis and ground-truth elements) format framework. In: 2010 20th International Conference on Pattern Recognition, pp. 257–260 (2010)
4. Tesseract OCR. https://github.com/tesseract-ocr/tesseract#tesseract-ocr
5. Recogniform Technologies. Recogniform Layout Analysis SDK. https://www.recogniform.com/layout-analysis.htm
6. Bhowmik, S., Kundu, S., Sarkar, R.: BINYAS: a complex document layout analysis system. Multimed. Tools Appl., 8471–8504 (2020). https://doi.org/10.1007/s11042-020-09832-3

Chapter 6
Summary

Abstract Digitization of paper documents is necessary to achieve compact and lossless storage, easy maintenance, efficient retrieval, fast transmission, and easy access. To do that analysis of documents' layout is necessary. The objective of DLA system is to extract the document structure both geometrical and logical. A typical DLA system possesses a pre-processing stage and layout analysis stage. Pre-processing transforms the input image to a specific form, suitable for further analysis. Layout analysis involves region segmentation to extract structural information and classification to get the semantic labelling of the extracted regions. This chapter summarizes the previous five chapters describing various stages related to DLA and also gives the future direction.

Keywords DLA · Summary · Pre-processing · ABCD · BINYAS · Region segmentation · Region classification

Digitization of paper documents is necessary to achieve compact and lossless storage, easy maintenance, efficient retrieval, fast transmission, and easy access. However, mere electronic conversion would not be of pronounced help for properly preserving these documents as well as automatic information retrieval, unless we provide a system for efficiently analysing the layout of these documents.

Documents possess explicit structure, which can be segregated into a hierarchy of physical modules, such as pages, columns, paragraphs, text lines, words, tables, figures, etc., or a hierarchy of logical modules, such as titles, authors, affiliations, abstracts, sections, etc.; or both. The special arrangement of these physical modules and their relationship in a document can be considered as the layout or structure of this document. The objective of DLA is to detect these physical modules present in an input document image to facilitate effective indexing and automatic information retrieval. A typical DLA system possesses a pre-processing stage and layout analysis stage. However, this may vary based on the type of document under consideration. Few methods also perform post-processing.

Pre-processing transforms the input image to a specific form, suitable for further analysis. Documents often suffer from many types of degradations for various reasons. Presence of noise may affect the quality of the final output negatively.

S. Bhowmik, *Document Layout Analysis*, SpringerBriefs in Computer Science,
https://doi.org/10.1007/978-981-99-4277-0_6

Therefore, pre-processing becomes necessary for such situation. Among all, binarization is commonly used as a pre-processing step before DLA. Binarization separates foreground pixels from the background pixels, so that every pixel will contain either 0 or 1 as intensity value. Binarization methods can be *threshold based*, *optimization base*, and *classification based* [1]. Threshold based methods compute a threshold value on pixel intensity either at the global level or at the local level to identify each pixel as foreground or background. Optimization based methods mostly use CRF for segregating the image pixels as foreground or background. Classification based methods follow supervised and unsupervised learning for recognizing image pixels as foreground or background. Deep learning based methods, developed for document image binarization also come under the purview of this category. In last few years, deep learning based methods turn out to be the state-of-the-art. For the evaluation of binarization methods many standard datasets and metrices are made available publicly through various competitions. These competitions are arranged periodically to offers a standard platform for evaluating methods against new set of images with more challenges [2].

After pre-processing, the next step is layout analysis. In this step, first, the input document image is segmented in various regions and then the segmented regions are classified as per their functional roll. It is often very difficult to separate the region segmentation part from the region classification part. Here, for better understanding, region segmentation methods are grouped based on the exact elements of the input documents the methods consider to start the analysis for segmentation. Accordingly, four groups are considered: *Pixel analysis-based methods*, *Connected component analysis-based methods*, *Local region analysis-based methods*, and *Hybrid methods*. Pixel analysis based methods, consider foreground, or background or both at the initial step to segment the input image. From the earliest methods to recent deep learning based methods come under the purview of this category. Methods from this category are very expensive in terms of processing time. Besides deep learning based methods require huge labelled data for training. Therefore, absence of sufficient data may affect the performance of these deep methods. However, recently, many large annotated datasets are made available for the training of these method. These methods are also found very effective for historical documents. Connected component based methods classify the connected components present in the input document image to obtain the segmented regions. CC analysis-based methods can handle the skewness of the documents if the inter-line gap is smaller than the inter-paragraph gap which may not be the case for all the documents [3]. Most of the recent methods developed based on the object detection based models are belong to the Local region analysis based methods category [4]. These methods have also obtained very impressive result. However, the performance of these techniques is highly sensitive to the process of local region extraction. Hybrid methods combine two approaches. For example, these may perform both pixel and component-level analysis to segment an input document image. Many of the recently proposed techniques follow this approach [3, 5–7]. Although most of these methods are not skew-invariant, for non-skewed documents, these are turned out to be the most efficient methods, even for document images with non-Manhattan layout. During segmentation, most of the recent deep learning based methods directly classify the regions in various semantic classes. However, most of the non-deep learning based methods, initially, identify the regions as text or non-text [8]. After that the text regions are farther classified in various semantic classes like, title, header, footer, drop-capital, authors, paragraphs, etc. [9, 10]. Non-texts are further classified as image, figure or charts, table, separators, etc. [11]. Tables [12] and charts [13] are further processed to extract information.

For better understanding of the reader, two case studies are given in Chap. 5. One is for historical documents and another is for contemporary documents. In contrast to the contemporary documents, the layouts of the historical printed documents are relatively constrained. However, historical documents suffer from many types of quality level degradation like uneven illumination, faded ink, clutter, and artefacts, such as dark patches, bleed through, and creases. For the layout analysis of these documents, a method named ABCD is discussed. As historical documents have huge intensity variation, binarization may produce many tiny discrete components. These tiny components often appear because the actual components get fragmented due to intensity level variation. This, in turn, produces the wrong boundary of the original component. Therefore, in addition to binarization, a distance transform-based technique has been devised to preserve the actual boundary of original components and to merge the fragmented tiny components to their source components. The whole method for segmenting historical printed documents is discussed in Chap. 5. In this chapter, a method name BINYAS [11] developed for the layout analysis of contemporary documents is also discussed. BINYAS performs CC analysis for text non-text separation as well as for non-text classification, and for text region generation or segmentation, it uses an adaptive and iterative morphology based method. Unlike the other methods, which use mathematical morphology for text region segmentation, it uses a rotating structuring element. The dimensions of which are set dynamically at each iteration. This method also estimates component level variation of the input image to automatically determine the next steps to be performed so that the unnecessary errors can be avoided and the computation time can be lessened. Here, the non-texts are classified as separator, table, graphic, chart, image, and inverted text. BINYAS can efficiently deal with Non-Manhattan layout along with the simpler layouts (i.e., Rectangular and Manhattan). Up to some extent, this method also works for documents with Overlapping layout.

Recently, deep learning based methods for DLA become dominant in the DLA field. As said earlier, deep model based methods require huge training data and training time, the absence of labelled data may affect their final outcome. Besides, these methods often require some post processing to fine tune their final outcome.

Therefore, more efforts are required to advance such methods in order to develop a generalize method which can deal with wide range of layout classes.

References

1. Tensmeyer, C., Martinez, T.: Historical document image binarization: a review. SN Comput. Sci. **1**(3), 173 (2020)
2. Pratikakis, I., Zagoris, K., Karagiannis, X., Tsochatzidis, L., Mondal, T., Marthot-Santaniello, I.: ICDAR 2019 competition on document image binarization (DIBCO 2019). In: International Conference on Document Analysis and Recognition (ICDAR), pp. 1547–1556 (2019)
3. Vasilopoulos, N., Kavallieratou, E.: Complex layout analysis based on contour classification and morphological operations. Eng. Appl. Artif. Intell. **65**, 220–229 (2017)
4. Zhong, X., Tang, J., Yepes, A.J.: Publaynet: largest dataset ever for document layout analysis. In: 2019 International Conference on Document Analysis and Recognition (ICDAR), pp. 1015–1022 (2019)
5. Antonacopoulos, A., Clausner, C., Papadopoulos, C., Pletschacher, S.: ICDAR2015 competition on recognition of documents with complex layouts-RDCL2015. In: 2015 13th International Conference on Document Analysis and Recognition (ICDAR), pp. 1151–1155 (2015)
6. Clausner, C., Antonacopoulos, A., Pletschacher, S.: ICDAR2017 competition on recognition of documents with complex layouts—RDCL2017. In: Proceedings of the International Conference on Document Analysis and Recognition, ICDAR, vol. 1, pp. 1404–1410 (2017). https://doi.org/10.1109/ICDAR.2017.229
7. Li, M., et al.: DocBank: A Benchmark Dataset for Document Layout Analysis. arXiv Prepr. arXiv2006.01038 (2020)
8. Bhowmik, S., Sarkar, R., Nasipuri, M., Doermann, D.: Text and non-text separation in offline document images: a survey. Int. J. Doc. Anal. Recognit. **21**(1–2), 1–20 (2018)
9. Tran, T.A., Na, I.S., Kim, S.H.: Page segmentation using minimum homogeneity algorithm and adaptive mathematical morphology. Int. J. Doc. Anal. Recognit. **19**(3), 191–209 (2016)
10. Tran, T.A., Oh, K., Na, I.-S., Lee, G.-S., Yang, H.-J., Kim, S.-H.: A robust system for document layout analysis using multilevel homogeneity structure. Expert Syst. Appl. **85**, 99–113 (2017)
11. Bhowmik, S., Kundu, S., Sarkar, R.: BINYAS: a complex document layout analysis system. Multimed. Tools Appl. **80**, 8471–8504 (2020). https://doi.org/10.1007/s11042-020-09832-3
12. Embley, D.W., Hurst, M., Lopresti, D., Nagy, G.: Table-processing paradigms: a research survey. Int. J. Doc. Anal. Recognit. **8**, 66–86 (2006)
13. Davila, K., Setlur, S., Doermann, D., Kota, B.U., Govindaraju, V.: Chart mining: a survey of methods for automated chart analysis. IEEE Trans. Pattern Anal. Mach. Intell. **43**(11), 3799–3819 (2020)

Printed in the United States
by Baker & Taylor Publisher Services